*Biotechnology
and
Food Quality*

Biotechnology and Food Quality

(PROCEEDINGS OF THE
FIRST INTERNATIONAL SYMPOSIUM)

Editors

Shain-dow Kung

Center for Agricultural Biotechnology
Maryland Biotechnology Institute
University of Maryland

Donald D. Bills

Agricultural Research Service
United States Department of Agriculture

Ralph Quatrano

E.I. du Pont De Nemours & Co., Inc.

Butterworths
Boston London Singapore Sydney Toronto Wellington

Copyright © 1989 by Butterworth Publishers, a division of Reed Publishing (USA) Inc. All rights reserved.

No part of this publication may be reproduced, stored in a retrieval system, or transmitted, in any form or by any means, electronic, mechanical, photocopying, recording, or otherwise, without the prior written permission of the publisher.

Library of Congress Cataloging-in-Publication Data

Biotechnology and Food Quality
    Proceedings of the First International Symposium on Biotechnology and Food Quality, 10/17-19/88, University of Maryland.
    Includes bibliographical references.
    1. Food--Biotechnology--Congresses. 2. Food--Quality --Congresses. I. Kung, Shain-dow. II. Bills, Donald D. 1932-   . III. Quatrano, Ralph. IV. Title.
TP248.65.F66I57 1988   664'.07   89-17345
ISBN 0-409-90222-5

British Library Cataloguing in Publication Data

Biotechnology and Food Quality
  1. Food technology. Applications of biotechnology
  I. Title II. Kung, Shain-dow III. Bills, Donald D.
IIII. Quatrano, Ralph V. Biotechnology and Food Quality
664

ISBN 0-409-90222-5

Butterworth Publishers
80 Montvale Avenue
Stoneham, MA 02180

10 9 8 7 6 5 4 3 2 1

Printed in the United States of America

CONTENTS

Agricultural Biotechnology

Agricultural Biotechnology: The Benefits for Developed and Developing Countries
Rita R. Colwell — 3

Food Quality Education
Alicia Löffler and Fergus Clydesdale — 9

Food Quality, Biotechnology, and the Food Company
M. Allen Stevens — 27

Making Technology Transfer Work
J. G. Ling — 45

Biotechnology: Regulatory Considerations
Kenneth A. Gilles — 51

Biology: Now Is the Time
Ralph W. F. Hardy — 57

Genetic Engineering and Food Quality

Characterization and Modification of Maize Storage Proteins
Brian A. Larkins, John C. Wallace, Craig R. Lending, Gad Galili and Evelynn E. Kawata — 67

Genetic Modification of Traits of Interest to Consumers and Processors
David A. Evans — 83

Omega-3 Fatty Acid Improvements in Plants
Scott Bingham, David Kyle and Richard Radmer   103

Enhancing Meat Quality with Somatotropin (ST): Potentials, Limitations, and Future Research Needs
Norman C. Steele, Roger G. Campbell, Thomas J. Caperna and Morse B. Solomon   115

Summary
Paul H. Tomasek   137

Molecular Components of Food

Cell Wall Dynamics
Kenneth C. Gross   143

Prospects for the Use of Genetic Engineering in the Manipulation of Ethylene Biosynthesis and Action in Higher Plants
Anthony B. Bleecker   159

Tomato Fruit Polygalacturonase: Gene Regulation and Enzyme Function
Alan B. Bennett, Dean DellaPenna, Robert L. Fischer, James Giovannoni and James E. Lincoln   167

Molecular Interactions of Contractile Proteins
Marion L. Greaser, Keh-Ming Pan, Jeffrey D. Fritz, Laura J. Mundschau and Peter H. Cooke   181

Molecular Components of Food: Repartitioning Agents or Hormones
Stephen B. Smith   201

Summary
C. J. Brady   213

Evaluation of Food Quality

Model Non-Isotopic Hybridization Systems for Detection of
Foodborne Bacteria: Preliminary Results and Future Prospects
S. W. Chan, S. Wilson, H-Y Hsu, W. King, D. H. Halbert and J. D. Klinger — 219

Use of RFLPs Analysis to Improve Food Quality
Tim Helentjaris — 239

Hybridoma Technology: The Golden Age and Beyond
R. A. Goldsby — 253

Summary
Richard Holsten — 269

Bioprocessing

Bioprocessing of Meats
L. Leistner and F.-K Lücke — 273

Genetic Modification of Enzymes Used in Food Processing
Bhav P. Sharma — 287

Impact of Biotechnology on Vegetable Processing
N. D. Addy — 307

Genetic Engineering of Lactic Starter Cultures
Larry L. McKay — 317

Production of Food Additives and Food Processing
Enzymes by Recombinant DNA Technology
*Jiunu S. Lai, Jar-How Lee, Shau-Ping Lei, Yun-Long Lin,
Joachim L. Weickmann and Lindley C. Blair* — 337

Summary
Todd Klaenhammer — 355

PREFACE

Biotechnology has great potential for quantitatively and qualitatively modifying agriculture. Quantitatively, we anticipate more efficient production, processing, and distribution of food and fiber. For example, genetically altered plants and animals can be made to be more resistant to diseases and to require lower inputs of fertilizer or feed; enzymes and microorganisms are being genetically designed to carry out food processing operations more efficiently; and new tissue culture techniques are revolutionizing the way that many plants are propagated. Changes such as these will increase the availability and decrease the cost of agricultural products without noticeably changing the nature of the final product. Consumers will be aware of quantitative changes only to the extent that retail prices are changed. On the other hand, qualitative changes that bring about desirable or undesirable changes in food quality will be highly visible.

The term *food quality* generally refers to the sensory properties of a food—appearance, taste, and texture, the characteristics that are detected directly by the human senses. For example, color, size, flavor, and crispness of an apple determine its acceptability and value, and perception of these attributes requires the senses of sight, smell, taste, touch, and hearing coupled with the resistance of the apple to the jaw and teeth. Nutritional quality is usually considered separately as is food safety, and these will be the topics of subsequent symposia in this series. Changes in food quality result from changes in the physical structure or chemical composition of a food. Slow changes have been brought about in the past through selection and breeding of plants and animals and through the development of food technology. Biotechnology offers a means for rapid and radical changes.

In affluent countries, superior quality is the key marketing advantage for a food product. Consumers are willing to pay a premium price for a food that has high sensory appeal, despite the fact that it may be nutritionally equal or even inferior to a competitive product. Attitudes about the attributes that constitute high quality are changing and are being modified more and more by other factors, such as perceived nutritional value and safety. The ideal contemporary product in the eyes of a modern consumer might be labeled "Light" (implying that it is low in calories), "High in Fiber" (implying a nutritional benefit), "Natural" (implying that competitive products that are not labeled "Natural" contain manmade, inferior ingredients), "Low Sodium" (implying that the product will help to prevent or control hypertension), and perhaps even "Organic" (implying that the food is safer because it was grown without the use of manmade fertilizers and chemicals). Yet in addition to all of these constraints, the ideal product also needs to be labeled "Gourmet" (implying that the food has exquisite flavor) and "Rich" (implying that it has high appetite satisfaction value). While the above is partly facetious, and knowledgeable consumers are not deceived by misleading implications, there is no question that the public strongly desires enhanced food quality and is willing to pay for it.

Consumers are receptive to improvements in food quality, but the application of biotechnology may encounter resistance if public concerns are ignored. Consider how the consumer attitude about the use of ionizing radiation will inhibit the development of an irradiated food industry in the United States. Reservations about the safety of the products and the false belief held by some that irradiated foods are radioactive will be difficult to overcome. Public involvement and education will be necessary to avoid unwarranted prejudices about foods that have been produced or altered by biotechnological means.

Biotechnology will influence food quality in many ways. The raw commodities themselves may be changed to improve their characteristics for further handling and processing, and it is important to recognize that inadvertent changes in quality may occur when changes that were intended to be only quantitative were sought. Processing of standard commodities may involve biotechnology to achieve modifications of the quality of the array of flavors, enzymes, and other food additives that will be useful in product formulation. The development of biological sensors and analytical procedures for measuring and controlling the quality of food products will provide another means of quality improvement. Obviously, many different types of expertise will be involved in the application of biotechnology to the improvement of food quality.

A central purpose of this symposium was to bring together scientists with interests and specific knowledge in the diverse areas that constitute the broad subject matter covered by the title of this book. Our intent was to promote communication and mutual appreciation of roles among molecular biologists, plant and animal physiologists, biochemists, food scientists, and others who are directly or indirectly involved in the improvement of food quality. We believe that the purpose of the symposium was fulfilled, and most of the credit must be given to the excellent speakers, who provided clearly phrased insights from their diverse positions of expertise. The remainder of the credit must be divided between the members of the Program Committee, who had the fine judgement to assemble such well-qualified speakers, and the members of the Local Arrangements Committee, who provided a favorable environment for our meeting.

We are indebted to the University of Maryland, the Agricultural Research Service of the USDA, and the DuPont Company, who sponsored the symposium and provided financial support and other facilities and services that contributed to the success of the First International Symposium on Biotechnology and Food Quality. This is a good example to illustrate a successful and mutually beneficial cooperation between a public university, a federal agency, and private industry.

Shain-dow Kung, *University of Maryland*
Donald D. Bills, *Agricultural Research Service, USDA*
Ralph Quatrano, *E.I. du Pont de Nemours & Co., Inc.*

AGRICULTURAL BIOTECHNOLOGY

Agricultural Biotechnology: The Benefits for Developed and Developing Countries

Rita R. Colwell
Maryland Biotechnology Institute
Department of Microbiology
University of Maryland
College Park, MD 20742, USA

Introduction

Rapid developments in medical biotechnology have already yielded products of societal value, the best examples of which are pharmacologically useful compounds and vaccines produced by genetic engineering. For the Third World, the greatest benefits of biotechnology will most likely come from agricultural biotechnology, notably in food production and protection applications. Pest control, disease resistance, extension of shelf life of agricultural products, and increased nutritional value of food crops will not only offer self-sufficiency for developing nations but also social and economic stability.

Biotechnological advances in food production should be based on international cooperation, since the introduction of genetically engineered food crops cannot be restricted by geography. Biological systems are notoriously disrespectful of national borders. The benefits of agricultural biotechnology in crop protection and food production, because of their enormous significance for all of humankind, should be shared among all countries—developed and developing alike. This is not to say that the economic rewards should not accrue to those who make the discoveries and provide the creative flight of intellect, but that international cooperation in research and development can effectively transfer the technology.

The introduction of field trials in both developed and developing countries simultaneously will provide the scientific data needed and also avoid ethical dilemmas, which may be posed either in the restriction of biotechnology on the basis of wealth and power to developed countries, or in the use of developing countries as merely the sites for field experiments. Cooperation in research and development at the international level can, and should be, a component of agricultural biotechnology.

Genes coding for useful traits, resistance to disease or insects, for example, have been transferred to crop varieties from non-cultivated plants, but recombinant DNA methods that are now available have greatly extended the sources of genetic information for transfer, including sources outside the Plant Kingdom. Research into selection for high-yield crops has been underway since the early 50's. In fact, during the 1960s, wheat and rice varieties were selected in Mexico and the Philippines from which new strains were used in other parts of the world. It is firmly established that they contributed to a significant increase in agricultural yields in Asian and Latin American countries.

The expression "Green Revolution" was coined to describe the increased agricultural production achieved in the developing countries by means of the new varieties, especially wheat and rice (Sasson, 1986). The unfortunate aspect of the Green Revolution was that cultivation of crops required extensive use of pesticides, irrigation and fertilization. From cross-breeding of the new varieties and hardy local breeds, well-adapted cultivars giving better yields were obtained. Millet and sorghum, triticale, maize, and several luguminous plant species, as well as wheat and rice, were included in Green Revolution research.

The exciting aspect of the Green Revolution is that within approximately ten years more than fifty percent of the surface of corn-growing land and thirty percent of the rice-growing areas in developing countries were sown with high-yield cereal varieties. With sufficient irrigation, fertilizer and weed killer, the yield is two or three times greater than that of traditional varieties. Introduction of new varieties of wheat to India in 1966 resulted in a doubling of Indian wheat production within a decade, reaching approximately 23 million tons in 1970, and eventually 33 million tons by 1980. India was catapulted from the second largest cereal importer in 1966 to self-sufficiency by the end of the 70's (Sasson, 1986).

In addition, where sufficient water was available in some areas of Asia, a shortening of the growing period was achieved so that at least two and often three crops could be harvested in a given year.

Those who benefited from the Green Revolution, unfortunately, were generally the wealthier farmers of a few of the developing countries, with those countries of Africa south of the Sahara not really affected significantly. Indeed, the new varieties of wheat and rice were not introduced with the same speed and intensity there as in Asia. Furthermore, the availability of irrigation and adequate fertilizer supplies, not as readily available in Africa, were very important in Asia as a factor in the Green Revolution (Sasson, 1986).

Gene transfer in crop improvement has been summarized, and documented, including the long history of gene transfer, carried out by plant breeders, between plant species and even between plants from different genera (Goodman et al., 1987). The use of recombinant DNA-based methods for gene transfer to plants and examples of the potential of this aspect of plant biotechnology in future crop improvement are well documented in the literature. Plant breeding is, of course, not new, with its scientific beginning traceable back to the 19th century, when discoveries of how plant traits are inherited were made. The early developments included transfer and reassortment of large numbers of genes in heterogeneous cultivated populations. Breeders extended the search for new genetic variation to the entire crop species, including non-cultivated populations. From these gene exchanges, modern cultivated varieties originated (Goodman et al., 1987). Thus, crop improvement involving interspecific and even inter-generic gene transfer is not new. Gene transfer by recombinant DNA is simply the latest in a long history of increasingly more powerful methods for crop improvement.

Genetic engineering of plants has not occurred without controversy. The issue attracting the greatest attention is release of genetically engineered organisms to the environment. Regulation of the uses of recombinant DNA technology and the fear of the unusual power of this technology, especially the uncertainty it creates in predicting the behavior of organisms genetically modified in new ways, are topics of intense debate. Experience with chemicals in the environment over the past fifty years reaffirm the conclusion that genetically modified organisms should be introduced into the environment only after extensive study and deliberation.

What has not been publicly debated is the use of biotechnology internationally. It can be argued whether biotechnological-based agriculture is intrinsically safer than traditional agriculture. However, hungry people need food, and therefore making agriculture more efficient should benefit those most in need. Two corollaries derive. One is that more efficient production generally signifies cheaper products and leads to a "squeezing out" of the inefficient producer who is too often the small or medium-scale farmer in the developing countries (Bunders and Boon, 1987). Traditional agricultural products are being replaced with other products that in some cases were developed using biotechnology, e.g. high fructose corn syrup and aspartane replacing cane sugar (Hobbelink and Ruivenkamp, 1987).

The strongest argument, initially, for biotechnology was that it would benefit poor farmers in developing countries. Experience based on the Green Revolution showed that when new developments increased total world food production, the result was further marginalization of resource-poor producers. In fact, the new technologies turned out to be inappropriate for small scale agriculture and the result was decreased produce prices. In fact, the decrease was spectacular in many cases (Stolp and Bunders, 1989).

Attempts have been made to organize effective communication between biotechnologists and groups involved in world food production, motivated by the belief that agricultural biotechnology could be oriented toward alternative directions. Questions were raised as to whether biotechnology could be effectively used to decrease the problems of small-scale farmers and pollution. The main question, of course, was whether biotechnology research could be oriented toward benefiting developing countries. The opportunity to talk face to face with people from a large spectrum of backgrounds will allow needed debate on how recent developments in agriculture can deal with the problems of marginalization of small-scale farmers, whether farmers will continue to be producers of raw materials only, and concerns about the environment involved in application of biotechnology to agriculture.

Thus, international workshops dealing with these issues can be highly beneficial. The question is whether biotechnology can cause a second revolution, a "Gene Revolution," to follow the Green Revolution. The negative aspects of the Green Revolution, as stated above, were that large quantities of fertilizers were needed to give high yields. Also, the susceptibility to diseases of mono-crop cultures have raised some serious questions about the use of biotechnology, especially in developing countries. It has been stated that biotechnology must lead to self-reliance, which implies that the process should profit from local resources and fit in with the norms and values of the local population (Stolp and Bunders, 1989).

If biotechnology is to have a positive impact on Third World countries, it simply must contribute to the self-reliance of those countries. Social and cultural factors must be taken into account, as well as technical factors, so that the success of biotechnology and its ancillary technolgoies can be assured.

Robertson and Sakina (1989) suggest the best solution for Africa, as one example, appears to be an increase in the yield of staple food crops, such as cassava and maize. They point out that peasant farmers in Zimbabwe achieved an average cassava yield of just under four tons per hector. But poor soils, lack of irrigation, and chronically low and unreliable rainfall resulted in a maize crop failure. Nevertheless, high hopes of biotechnology remain because of the belief that it is intrinsically transferable. The analogy of transferring a Mercedes-Benz factory to Africa, which would be desirable but almost impossible because of the lack in Africa of a critical mass of engineering expertise, doesn't apply: a molecular biology laboratory takes only a container full of equipment and a busload of scientists. The result of the biotechnology transfer could be production of a new cultivar of maize or a vaccine against sleeping sickness. Thus, carefully choosing the biotechnology and focusing it on transforming the quality of life in the Third World is a noble, and achievable aim.

It is necessary to be realistic about what is needed to support an appropriate biotechnology in the Third World. Obviously, the first premise is to ensure that negative effects of biotechnology are minimized. In developing countries lifestyles are much more biological than in developed nations, and a much greater proportion of the population is involved directly in food production; thus, there is less opportunity for chemical and physical intervention in agriculture or for medical and veterinary applications. Clearly, where biotechnology can provide an appropriate solution is the crux of the matter, as well as whether the developed countries are willing to divert some technological and financial resources towards the needs of developing nations (Bunders et al., 1989).

Introduction of novel, genetically engineered crops into countries where environmental regulations are relatively less stringent, because the benefit seemingly outweighs the risks, should be undertaken only with adherence to internationally accepted standards for introduction of genetically engineered organisms to the environment. Obviously, an effort should be made to establish an international code for the introduction of genetically engineered organisms, whether in the developed or developing countries.

Together scientists and public interest groups can formulate directions for research in biotechnology. Together they can convince governments that the development of policies stimulating biotechnology should be aimed at sustainable development in developing countries. Such cooperation is not only feasible but vital if biotechnology is to thrive; if agricultural biotechnology to fulfull its promise.

References

Bunders, J. and Boon, L.J. 1987. Biotechnology. Vol. I, (Neijsser, O.M., van der Meer, R.R., and Luyben, K. Ch.A.M., Eds.). Pgs. 351-354, Elsevier Science Publishers.

Bunders, J., Broerse, J. and Stolpe, A. 1989. Necessary, robust and supportable: the requirements of appropriate biotechnology. Tibtech January 1989, Vol. 7, One World Biotechnology, pages S16-S24.

Goodman, R.M., Hauptley, H., Crossway, A. and Kanuf, V.C. 1987. Gene Transfer and Crop Improvement, SCIENCE, Vol. 236:48-54.

Hobbelink, H. and Ruivenkamp, G. 1987. Biotechnology in the Third World: A demasque of a new promise, international coalition for development action.

Robertson, A.I. and Sakina, K.E. 1989. A slice of reality from Africa. Tibtech, January 1989, Volume 7, One World Biotechnology, pages S14-S15.

Sasson, A. Biofutur/UNESCO, Paris, 1986, "Quelles biotechnologies pour les pays en developpement?"

Stolp, A. and Bunders, J. 1989. Biotechnology: wedge or bridge?. Tibtech, January 1989, Vol. 7, One World Biotechnology, pages S2-S4.

Food Quality Education

Alicia Löffler and Fergus Clydesdale
Department of Food Science and Nutrition
University of Massachusetts
Amherst, MA 01003, USA

The long-term acceptability of a new food product ultimately depends on its quality. Thus, there has been a great deal of interest in the development of technologies for the control, improvement, and prediction of food quality. Unfortunately, the field of food quality has tradionally been viewed as routine, and perhaps as more art than science. This traditional view is now being challenged, in part because of increasing recognition of advances brought about through biotechnology.

Food quality measures must provide the speed and accuracy essential for the new type of flexible mass production we will see in the future, and this is where biotechnology has found a perfect "niche": in the interdisciplinary development of quality biotechnology. This integration of biotechnology and food quality is stimulating a cascade of scientific and technical advances for the measurement and control of many food quality parameters. Consequently, food quality is now being perceived as a highly scientific and technical field and is becoming more attractive to basic science students.

This paper discusses the impact that the marriage of food quality and biotechnology will have on academic institutions, with respect to both education and research. We will expand on the available tactics to implement food quality-biotechnology in the education of food scientists, the contribution of universities in the recent research of developments of food quality, and the type of research in food quality-biotechnology that best fits the university system.

Introduction

Projections of future world food supply seem to concentrate solely on increased yields, food production, and feed ratios. As important as these are, quality is the pivotal parameter in any future scenario involving more and better foods. Kramer and Twigg (1966) defined "quality" some years ago as:

> "the composite of characteristics that differentiate individual units of a product and have significance in determining the degree of acceptability of that unit by the buyer. Thus the over-all quality of a food product not only may be, but should be, ana-

lyzed for its components attributes, each of which should be measured and controlled independently. The more completely and precisely a specific attribute can be defined, the more probable the attainment of a satisfactory instrumental method for its measurement. Whenever any work is performed on a food item which differentiates it from the mass of the product, be it more than a segregation according to its size, something has been added to its quality."

We believe it is obvious from this definition that quality will be pivotal in the future of the food industry and world food supply.

Unfortunately, as important as quality is, it has never enjoyed a prominent role in basic education. This does not mean to imply that education in quality does not have the same basic scientific discipline and rigor of other areas, but simply that its perception has not included the "romanticism" of many other scientific studies.

Scope and Future Role of Food Quality

In order to combat such perceptions it is necessary to provide a realistic and exciting role for food quality in the future. To do this, the following questions should be answered:

- What should quality encompass?
- How will it help to reshape the future of the food supply?
- What disciplines and tools are necessary to meet future goals?

The answer to the first question is that quality should be all-encompassing according to the definition given previously and to future projections. It should include both the measurement and development of ingredients and systems to increase and insure quality, in terms of sensory characteristics, safety, nutrition, and consumer demands.

The second question is of utmost importance not only for the production of food for an increasing world population but for the proper positioning of quality as a technological tool in the shaping of the U.S. economy. A new study by the Office of Technology Assessment (OTA) indicates that the American economy is being reshaped by three principal forces: new technologies, a rapid increase in foreign trade, and changing tastes and values (Norman, 1988). More importantly perhaps for this discussion is that the report suggests that:

> "the ability to respond quickly to changing concerns will place a premium on flexibility and should force corporations to move away from the centralized economic management and mass production that has dominated the industrial scene for much of the 20th century."

Clearly if the food industry is to meet this scenario, then quality will become even more important and will require a pool of expertise that will only be gained through new efforts on the part of the education system.

The third question hinges on the needs for meeting the goals of flexibility that allow for rapid change as well as innovative new products that meet consumer demands. This question is really the focal point for bringing together quality, biotechnology, and education because it is this combination that will allow the goals to be met. The increased use of biotechnology in the field of food quality will allow more rapid assessment technologies, new constituents, and the ability to utilize quality measures to predict sensory, safety, and nutritional quality after processing and storage.

Having established the triumvirate of "quality-education-biotechnology," it is necessary to develop a scheme for:

- how biotechnology is to be woven into the educational fabric of a food scientist, and
- the type of research in the quality-biotechnology area which best fits the university.

Before developing a conceptual framework for biotechnology in the Food Science curriculum, it is necessary to attempt to define it. A broad definition has been developed by the Cabinet Council Working Group on Biotechnology as "the application of biological systems to technical and industrial processes" (Miller and Young, 1987). More commonly, however, a narrower definition is used which refers to the modification of biological systems for their application in innovative processes. This definition encompasses the technologies responsible for vinegar, wine, and beer, which began 8,000 years ago, to the present molecular and genetic manipulations which when combined with advances in electronics and engineering have produced such sophisticated techniques as molecular modeling, and protein and metabolic engineering. Because of biotechnology's rapid evolution and dependence upon a wide range of scientific areas, it is difficult to fit it into any set mold or discipline. For example, ten years ago, the technique of cloning for product amplifications and transfer was thought of as biotechnology. Today, cloning is considered just a tool and has very little weight on its own. The same holds true for molecular modeling. It is thus futile to link any specific technique, set of techniques, or even disciplines to biotechnology since it is continuously changing and evolving.

The inability to specifically define and categorize biotechnology creates a new challenge for educators. Due to its nature, no professional body has undertaken the burden of standards and oversight for teaching biotechnology. Things move so quickly that it is very difficult at any given time to know the finite boundaries of a food education in food biotechnology. This raises the following unsolved questions: Is it possible to teach and train students in biotechnology knowing that the material being covered at the beginning of the semester might become somewhat obsolete at the end of that same semester?; Should single departments be responsible for covering the whole field of biotechnology?; and, What type of research in food biotechnology should exist in academic institutions?

The following discussion will attempt to address these issues under two headings, "Undergraduate Education" and "Graduate Education and Research."

Undergraduate Education

The competitive advantage of Food Science and Technology has been in large measure due to the United States university system. Schools have provided a pool of properly trained individuals that can teach, make further technological advances, and transfer these findings into commercial settings. This has been achieved in Food Science departments through rigorous training in traditional disciplines such as chemistry, microbiology, and engineering.

As biotechnological advances move from research to development and manufacturing, food industries are beginning to demand a new mix of biotechnological skills. Should universities be responsible for designing programs in food biotechnology to train students that could satisfy these demands?

As suggested before, biotechnology is rather a function than a discipline. It certainly draws from molecular biology, immunology, microbiology, cell biology, engineering, computer science, etc. Biotechnology itself is all, but none of the above. Restricting biotechnology to a department or an area such as Food Biotechnology, Microbial Biotechnology, etc., to conform to the boundaries of an academic discipline is dangerous. Most often, when this is done, the cross-disciplinary nature and rapid evolution of the field becomes fragile and incomplete. Moreover, the restricted number of faculty members of Food Science departments makes it impossible to cover all disciplines that biotechnology encompasses.

There is no doubt that the Food Science undergraduate curricula should retain the traditional courses in general biology, molecular and cellular biology, chemistry, physics, and microbiology which coincidentally also represent the backbone of biotechnology. More than ever, Food Science departments should be responsible for making sure that those courses are taught in a stimulating manner and that they incorporate the most recent developments.

In addition, departments of Food Science should take responsibility in training students to intelligently interact with a variety of specialized disciplines, to learn how to extract the innovative concepts from those specialized areas, and to creatively integrate these concepts and transfer them into new food technologies.

In recognition of these needs, most Food Science departments have appointed a faculty member to teach a course in "Food Biotechnology" which draws all the disparate disciplines together and points to the potential applications in food technology. It also presents an overview of the most recent biotechnological advances in the agricultural industry. Although such a course is usually very informative, it most often results in a limited exposure of the students to cross-disciplinary concepts.

An alternative approach would be to teach such a course in cooperation with not only other departments but with members of industries involved in biotechnology. This formula was introduced in 1986 by Tufts University. A series of courses at the senior and Masters level were designed with a consortium including more than 20 companies. This experiment appeared to be very successful and should be definitely explored in more detail by Food Science departments in the United States.

Perhaps courses such as these should be taught prior to the senior year so that the techniques learned can be incorporated into such courses as Food Analysis, Food Processing, and subsections of the courses which deal with "quality." Indeed, with the importance of quality, consideration should be given to a full course with the objective of teaching "technology transfer" of biotechnology to quality as well as principles of food quality itself. This would have several advantages. First, it would provide food quality with an "upscale" image to attract top students. Second, it would provide trained personnel to fill the needs in this field. Such a course could provide significant hands on-experience, an essential component of any such program.

Graduate Education and Research

The purpose of graduate school is to provide an environment where the student may learn to think and conduct independent research. In general, graduate students in Food Science learn to formulate and activate a research plan that will ultimately improve both food quality and production. The specific area of research in which they are trained is not as important as obtaining the basic tools for approaching a research problem. While undergraduate training in food biotechnology should stress the ability to interact with many disciplines, the graduate student should be able to become specialized in his/her research area. On the surface, this idea seems to contradict the very nature of biotechnology and many educators are questioning whether the traditional approach of training graduate students should be modified.

While the concept of training M.S. and Ph.D.'s in "food biotechnology" seems attractive and fashionable at first, it also presents some serious disadvantages. For example, it is impossible for an individual to become competent in all disciplines of biotechnology during the limited period necessary to obtain an M.S. or a Ph.D. Most likely, if this is attempted, the student would become a "generalist" unable to attack a research problem with any serious depth. Paradoxically, by training students as "generalists," we would also be damaging the field of biotechnology. As said before, biotechnology relies on the integration of very specialized and sophisticated advances in a variety of scientific disciplines. Graduate school is perhaps the most specialized and productive stage of a scientist's career; by creating "generalists" we would deprive the field of biotechnology of important pool of specialized scientists essential for its growth.

Therefore, it is not surprising to see that many major biotechnological advances occur in industrial environments where teams of scientists from diverse backgrounds are brought together for specific projects and dissolved when the projects are completed. A typical example would be the protein engineering teams assembled at different genetic engineering companies (e.g., Genentech) to work on the design of specific enzymes. These teams consist of people of diverse backgrounds such as mathematics, computer science, molecular biology, chemistry, and microbiology. This flexibility does not generally exist at universities, where faculty members usually have departmental loyalties. These faculty members are not hired to become part of a larger research project but rather for their ability to teach and conduct independent research. Consequently, research teams are very difficult to organize in academic environments. Does this mean that we should forget about the idea of conducting biotechnological research at Universities? Absolutely not. Instead, we need a rational program organized through both the industrial and academic sectors to identify general areas in food biotechnology in need of long-term research studies which are best suited to universities, where the expectations for immediate applicability are low.

In the remainder of this paper we will attempt to identify several general programs in food quality biotechnology which we believe could be best satisfied by universities. These programs are all in need of fundamental research with very long-term goals. Some of these programs are already in progress and they represent the path from raw agricultural products to the consumer.

Improvement of Agricultural Raw Materials

This program would include all manipulations of agricultural raw materials (animal and plants) directed to improve their functionality, nutritional value, or cost. There is an infinite number of important projects within such a program and all of them require basic research to increase our understanding of the molecular interactions involved in the physiology, metabolic regulations, and genetics of the raw materials to be manipulated. Some specific examples follow.

Improvement of the color characteristics of food plants
There is a great deal of interest in trying to preserve the green color of vegetables. All studies done so far seem to indicate that the preservation of green color ultimately depends on the ratio between chlorophyll a and chlorophyll b (Aronoff, 1966). Normally, this ratio is approximately 3:1. Unfortunately, chlorophyll a is converted to pheophytin 5-10 times faster than chlorophyll b (Clydesdale and Francis, 1976). Metabolic engineering techniques could be applied to green vegetables to reverse the ratios between chlorophyll a and b to increase color stability. However, we still lack the fundamental knowledge of the genetics and physiological regulation of these pigments required to achieve this goal.

Similarly, we are familiar with the efforts being made to increase the stability of plant anthocyanins used as food colorants not only to improve their quality but also to standarize their tinctorial power and increase their production. Unfortunately, the flavylium ring of plant anthocyanins is very reactive, making the pigment very prone to degradation due to pH, presence of sulfite, sulfur dioxide, ascorbic acid, amino acids, etc.

Recently, plant cell culture methods have been developed for the production of anthocyanins (Iker, 1987). We could potentially use metabolic engineering techniques to induce this plant cell culture to produce anthocyanins with phenyl or methyl groups at position 4 which is known to make the pigment stable to all the agents listed above (Clydesdale and Francis, 1976). Again, to achieve this goal we still need to increase our understanding of the regulatory genetics for the synthesis of plant anthocyanins.

Improvement of the functionality and resistance of plants
In the same way as it is now possible to design transgenic animals with altered fat/protein ratios, it could be possible to manipulate tomato plants so as to increase their solids content and thereby reduce the transportation and heat processing costs. It has been estimated that if the solid content of tomatoes could be increased by just 1%, the food industry could save $80 million dollars/year (Agric. Genet. Report, 1983). We could also increase the amounts of unsaturated fatty acids in the membrane phospholipids of tomatoes to increase their resistance to chilling injury. Obviously, these projects are still in need of very fundamental studies on the molecular biology of the tomato plant.

This program would also include all the current studies aimed at increasing the resistance of plants to pests and thus minimizing the use of potentially expensive and dangerous pesticides. Typical examples are the development of transgenic tomato and tobacco plants resistant to lepidoptera larvae (Fischoff et al., 1987; Crawford, 1986) and the development of transgenic walnut trees that could be resistant to codling moths and navel orange worms (McGranaham et al., 1988). Eventually, it will be also possible to reduce the use of fertilizers by developing nitrogen-fixing properties in nonleguminous plants with *Rhizobium*. Research in this area has been very active and we should expect having nitrogen-fixing plants within the next 10 years (McCormick, 1988). The first step towards this goal occurred recently on March 10, 1988, when BioTechnica International (Cambridge, MA) received the first U.S. permission to field test recombinant *Rhizobium meliloti* that lives on the roots of alfalfa plants.

Improvement of the nutritional content of plant raw materials
Genetic engineering techniques could be applied to a variety of plants to improve their nutritional value. For example, one could design a plant to synthesize specific amino acids normally deficient in that plant (e.g., a corn plant capable of synthesizing large amounts of methionine). Moreover, one could reduce the content of certain compounds that diminish the bioavailability of nutrients. For example, tannins reduce the bioavailability of vitamin B12 while oxalates, phytates, and fibers affect the absorption of Ca, Fe, and Zn (Teutonico, 1987; Clydesdale, 1988). Further research into the mechanism of enhancement of nutrient bioavailability could also be utilized via genetic

engineering. For example, recent studies have isolated a peptide from chicken muscle which may be, in part, responsible for the enhancing effect of meat on iron absorption (Slatkavitz and Clydesdale, 1988). This factor could potentially be engineered into plants so that iron absorption would be enhanced rather than inhibited. This would definitely help to resolve the 500 million cases of anemia worldwide. Not surprisingly, to achieve these goals we still need to increase our knowledge of the genetics and physiology of the plants involved.

Improvement of Processed Foods

This program would include all manipulations of processed foods directed at improving their functionality and overall quality including their preservation, safety, flavor, and nutrition. As in the case with raw materials, there is an infinite number of examples that could be listed under this program. Most of them are in great need of basic studies to increase the molecular and physico-chemical properties of the biomaterials being manipulated. Typical projects within this program follow.

Improvement of the functional properties of processed foods
For many years it was believed that the only way to modify food products to improve their functionality was through enzymatic treatment, the most classical examples being plastein reactions and the enzymatic treatment of meat with papain for tenderization. However, protein engineering is now revolutionizing the idea of product functionality since it is now possible to design food components with any desired catalytic, colligative, and functional properties for each processing step without the use of enzymatic hydrolysis. Obviously, since each component behaves differently each should be studied independently. Thus, a tremendous amount of fundamental research will be required before this technology becomes applicable.

Enzymatic treatments for product modification could also benefit enormously by advances in protein engineering. Food enzymes could be designed to be resistant to all sorts of harsh processing conditions or to be active only at some specific temperature, water activity, etc. For example, we could design food enzymes to pre-digest a product only during cooking temperatures.

Improvement of nutritional content or acceptability of processed foods
Historically, food additives such as fatty acids, antioxidants, antimicrobial agents, flavors, vitamins, etc., have been produced synthetically or by classical microbial fermentations. With the advent of plant tissue technology there has been a significant interest in the search for new, more efficient, or safer food additives. Table 1 shows a representative list of some of the many secondary metabolites that could be of interest to the food industry. For example, shikonine could be used as a "natural" dye and antimicrobial agent and is produced by *Lithospermum erythrorizen* (Shimomura et al., 1986), or tocopherols (vitamin E) is synthesized by blue-green algae (Knorr and Sinskey, 1987). There are yet many other potentially useful secondary metabolites that await to be explored.

Table 1. Representative list of secondary metabolites relevant to the food industry

Secondary metabolite	Source	Reference
Shikonin	*Lithospermum erythrorizen*	Shimomura et al., 1986
Thaumatin	West African shrub	Knorr and Sinskey, 1987
Hydroquinone	*Bifurcaria galapagensis*	Harvey, 1988
Acrylic acid	*Phaecystis poouchetti*	Harvey, 1988
Cytokinins	*Phaephyta*	Harvey, 1988
Quinoline alkaloids	*Cinchona legeriana*	Hamill et al., 1987
Betacyanin, betaxanthin	*Beta vulgaris*	Hamill et al., 1987
Tocopherols	Blue-green algae	Knorr and Sinskey, 1987
Vanilla, strawberry, raspberry, and grape flavor	Undifferentiated cell culture of vanilla, strawberry, raspberry, and green plants	Anon., 1987
Gibberellins	*Gibberella fujikuri*	Demain, 1984
Cyclosporin A	*Tolypocladium inflatum*	Demain, 1984
Pepstatin	*Streptomyces spp.*	Demain, 1984

Not only the production of "natural additives" could be improved by recent biotechnological advances but also their bioavailability. For example, the gene for thaumatin produced by a shrub in South Africa, which is used as a sweetener (being 2,500 X sweeter than sucrose) and as a flavor enhancer, has been recently incorporated in the genome of yeast (Anon., 1987). The sweetener is then synthesized in large amounts, encapsulated, and stabilized by the yeast. Using similar techniques, we could use yeast to encapsulate and stabilize a variety of vitamins, essential amino acids, minerals, etc. The yeast itself could be used as a food additive and, at the same time, it most likely would enjoy the acceptability of the health-conscious consumer. As well, the idea of introducing such a non-nutritive sweetener into plants offers some interesting perspectives on the possibilites of a vast array of high-quality, low-calorie foods.

The search for new microbial polysaccharides is an area of active research, particularly those that simulate the texture and mouth-feel of fat, but are not digestible and therefore are non-caloric. More interestingly, we could potentially use protein engineering technology to design food ingredients would be polymerized by stomach enzymes and this would have a reduced caloric content.

Improvement in the preservation of processed foods

Microbial fermentations have been used for many years for the preservation and stabilization of dairy, meat, fish, fruit, and vegetable products. Unfortunately, the efficiency of these fermentations is quite low and additional fundamental studies on the molecular biology of these organisms is still required to improve them. For example, the dairy industry is in need of starter cultures resistant to phage attack, capable of competing with indigenous organisms, and capable of resisting very acidic conditions. Moreover, these starter cultures could potentially carry genes that could improve the nutritional quality of the product being fermented. For example, it has been proposed that the genes responsible for the conversion of cholesterol to coprostonol in *Eubacterium* be transferred into lactic acid bacteria so as to produce cholesterol-free fermented dairy products (Harlander, 1987; Venema and Kok, 1987). This goal is very near completion.

Protein engineering techniques could also be applied to design food ingredients resistant to oxidizing agents, temperature, or microbial attack, thus increasing the shelf life of the food product.

Improvement in food safety and quality control
Detection technology represents the pinnacle of biotechnology today, the perfect marriage between molecular biology and microelectronics which has revolutionized the field of product safety and quality control. The current literature is very rich in examples that support this statement (Fitts, 1986; Klausner, 1987; Van Brunt, 1987). Most of them made use of either monoclonal antibody, DNA hybridization, or biosensor technologies. Very sensitive probes are now available for the detection of pathogens, contaminants, and to determine the freshness and optimal harvesting time of particular crops. Table 2 summarizes some of the available probes relevant to the food industry.

Unfortunately, most of the advances in detection technology have been made by the pharmaceutical industry and there is still an urgent need to accelerate its development in the food science area. For example, DNA hybridization technology is still in its infancy for many important food pathogens such as *Campylobacter spp.* The same is true with the development of monoclonal antibodies for a variety of food contaminants and for the development of on-line biosensors for food processes.

Most importantly, we are in need of bioprobes that could allow not only the detection of products within a food but also the prediction of the shelf life of that product or the bioavailability of its nutrients. For example, it could be possible to develop biosensors that could predict the bioavailability of particular minerals by using a cell-membrane receptor laid down on a piezoelectric crystal. The binding of the mineral to the membrane would be proportional to the decrease in the frequency of the crystal.

Recently, it has been demonstrated that IR/VIS reflectance spectroscopy could be successfully used for the measurement of iron hydrolysis, and potentially its bioavailability (Eyerman et al., 1987). Thus, it is easy to imagine in the near future the development of on-line IR/VIS probes for the prediction of iron bioavailability for different food products in the near future.

Similarly, it is well known that NMR could be used to monitor the metabolic state of microorganisms and plant cells (Shanks and Bailey, 1987). It could be then possible to design on-line NMR probes to monitor production of secondary metabolites in growing cultures not only of microorganisms, but also of plant and animal cells.

Table 2. Representative list of bioprobes relevant to the food industry

Function	Type	Reference
On-line determination of methanol	Enzyme electrode	Belghith et al., 1986
On-line determination of biomass and metabolic state	Fluorescence	Van Brunt, 1987
Determination of meat freshness	Monoamine oxidase membrane with O2 electrode	Karube, 1987
Determination of fish freshness	5'-nucleotidase membrane and nucleoside phosphorylase; xanthine oxidase membrane with O2	Karube, 1987
Determination of glucose and glutamate	Microchip	Karube, 1987
Detection of Staphylococcus endotoxin	Monoclonal antibodies	Klausner, 1987
Detection of aflotoxin B1	Monoclonal antibodies	Klausner, 1987
Detection of herbicides (chlorosulfuron; 2,4,5 T; and paraquat)	Monoclonal antibodies	Klausner, 1987
Detection of insecticides (diflubenzuron and parathion)	Monoclonal antibodies	Klausner, 1987
Detection of polychlorinated biphenyls	Monoclonal antibodies	Klausner, 1987
Detection of pathogens (C. botulinum, Y. enterocolitica, V. parahaemolyticus, V. cholera, E. coli, L. monocytogens)	Monoclonal antibodies	Harlander, 1987

Conclusion

There is no doubt that biotechnology will play a pivotal role in the future of food quality. In this discussion we attempted to establish the role that academic institutions should have in facilitating this integration between biotechnology and food quality. Two important considerations should be remembered:

First, to best educate future food scientists in food quality biotechnology we should maintain a rigorous and updated scientific curriculum. Active student interactions with other disciplines should be encouraged or demanded, perhaps through a series of courses of workshops involving the industrial community.

Second, research studies at universities should continue to be focused on long-term goals. However, these studies should respond to national priorities. We attempted to identify some of the projects that could respond to national needs in food quality biotechnology. However, it would be presumptuous of us to expect to have presented a complete list of programs. A thorough list could only be developed through a concerted effort between the academic and the industrial sector.

References

Agric. Genet. Report. 1983. Nov/Dec:7.
Anon. 1987. Growing flavors in the lab. Science Impact Letters 1:1.
Aronoff, S. 1966. In: The Chlorophylls. (Vernon, L.P. and Seeley, G.R., eds.) Academic Press, New York.
Belghith, H., Romene, J.L., Thomas, D. 1986. An enzyme electrode for on-line determination of ethanol top methanol. Biotech. Bioeng. 30: 1001.
Clydesdale, F.M. 1988. Mineral interactions in foods. In: Nutrient Interactions (Bodwell, C.E. and Erdman, J.W., eds.). Marcel Dekker Inc.-Inst. Food Technol. p. 73.
Clydesdale, F.M. and Francis, F.J. 1976. Pigments. In: Principles of Food Science. Part I (Fenema, O.R., ed.). Marcel Dekker, Inc., New York. p. 385.
Crawford, M. 1986. Test of tobacco containing bacterial gene approved. Science. 223: 1147.
Demain, A.L., 1984. New applications of microbial products. In: Biotechnology and Biological Frontiers (Abelson, P.H., ed.). The American Association for the Advancement of Science, Washington, D.C. p. 201.

Eyerman, L.S., Clydesdale, F.M., Huguenin, R., and Zajeicek, O.T. 1987. Characterization of solution properties of four iron sources in model systems by solubility studies and IR/VIS reflectance spectrophotometry. J. Food Sci. 52: 197.

Fischoff, D.A., Bowdish, K.S., Perelak, F.J., Marrone, P.G., McCormick, S.M., Niedermeyer, J.G., Dean, D.A., Kusano-Kretzmer, K., Mayer, E.J., Rochester, D.E., Rogers, S.G., Fraley, R.T. 1987. Insect tolerant transgenic tomato plants. Bio/Technology 5: 805.

Fitts, R. 1986. Biosensors for biological monitoring. In: Biotechnolgy in Food Processing (Labuza, H.S., ed.). Noyes Publ., Park Bridge, New Jersey p. 271.

Hamill, J.D., Parr, A.J., Rhodes, M.J.C., Rhodes, R.J., and Walton, N.J. 1987. New routes to plants secondary products. Bio/Technology. 5: 800.

Harlander, S.K. 1987. Biotechnology: emerging and expanding opportunities for the food industry. Nutrition Today. 22: 21.

Harvey, W. 1988. Cracking open on marine algae's biological treasure chest. Bio/Technology. 6: 488.

Iker, R. 1987. *In vitro* pigment production: an alternative color synthesis. Food Technol. 41: 70.

Karube, I. 1987. International Conference and Exhibition for Bio and Gene Technology, Biotech '87, Dusseldorf, West Germany, March 17-19.

Klausner, A. 1987. Immunoassays flourish in new markets. Bio/Tecnology. 5: 551.

Knorr, D., and Sinskey, A.J. 1987. Biotechnology in food production and processing. In: Biotechnology: The Renewable Frontier. (Koshland, D.E., ed.) AAAS. Washington, DC. p. 253.

Kramer, A. and Twigg, G.A. 1966. Fundamentals of Quality Control for the Food Industry. AVI Publ. Co., Westport, Connecticut.

McCormick, D. 1988. How biotechnology is dealing with nitrogen fixation. Bio/Technology. 6: 383.

McGranaham, G.H., Leslie, C.A., Uratsu, S.L., Martin, L.A., Dandekar, A.M. 1988. Agrobacterium mediated transformation of walnut somatic embryos regeneration of transgenic plants. Bio/Technology. 6: 800.

Miller, H.I. and Young, F.E. 1987. Isn't it about time we dispensed with biotechnology and genetic engineering? Bio/Technology. 5: 184.

Norman, C. 1988. Rethinking technology's role in economic change. Science. May 20: 977.

Shanks, J.V. and Bailey, J.E. 1987. Application of 31PNMR spectroscopy to fermentation studies. Paper #158c. 1987. AIchE Annual Meeting, November 15-20, New York.

Shimomura, K., Satake, M., and Kamada, H. 1986. Production of useful secondary metabolites by hairy roots transformed with Ri plasmid. In: Proc. VI International Congress of Plant Tissue and Cell Culture (Somers, D., Gegenbach, B.G., Biesboer, D.D., Hackett, W.P. and Green, C.E., eds.). University of Minnesota, Minn. p. 250.

Slatkavitz, C.A. and Clydesdale, F.M. 1988. Solubility of inorganic iron as affected by proteolytic digestion. Am. J. Clin. Nutr. 47: 487.

Teutonico, R. 1987. Impact of biotechnology on the nutritional quality of foods. In: Food Biotechnology, Knorr, I., ed.). Marcel Dekker, Inc. New York. p. 95.

Van Brunt, J. 1987. Biosensors for bioprocesses. Bio/Technology. 5: 435.

Venema, G. and Kok, J. 1987. Improving dairy starter cultures. TIBTECH 5: 144.

Food Quality, Biotechnology, and the Food Company

M. Allen Stevens
Campbell Soup Company
Camden, NJ 08103, USA

Success in the food business depends on knowing what the consumer wants and then satisfying that need with a high quality product at a competitive price. To deliver the consistent quality expected, food companies are restructuring their quality assurance programs according to the Japanese model. The challenge is to provide high-quality food to satisfy changing eating habits resulting from demographic and social changes. Consumers want foods that taste good, are safe and nutritious; demand is increasing for freshness, convenience and all natural. A food must have good taste and aroma or it will fail. There can be absolutely no compromise on product safety; adversely affecting any consumer's health will usually severely compromise a food company. People are increasingly aware that eating habits affect health and longevity. Biotechnology offers considerable potential to improve the quality of foods; several specific examples are considered.

Introduction

There are several major factors that are impacting food companies and affecting their concern regarding quality. Food companies are competing for the food dollar of a population which has slow growth. Companies with favorable cash flows, sometimes from the sale of tobacco, have gained control of major food companies, and large foreign corporations have assumed greater importance in the U.S. food business. Consumer demands are changing as life-styles change and as the importance of diet to health and longevity becomes clearer. The adage that you are what you eat has assumed considerable importance in food buying patterns. The newer adage that you become what you eat is gaining credence.

As a result of intense competition for the consumer's stomach and palate, food companies are highly motivated to discover what they want and to deliver it. They want to compete more effectively for the food dollar and plan to learn to capitalize on the hard lessons learned in the automotive and electronic industries at the hands of the Japanese. There is a strong desire to use quality to attract new customers and to gain market share.

Quality Assurance

Comparison of quality control procedures traditionally used by many companies in the U.S. with those so successfully used by Japanese companies will serve to illustrate major differences in philosophy (Table 1). To further emphasize the old (not too successful) and the newer quality assurance systems, a few comparisons are provided (Table 2).

Table 1. A comparison of unsuccessful U.S. strategies for quality assurance with those of successful Japanese methods

Item	U.S.	Japan
Quality policy	Various policies adopted, but generally consist of and may be limited to quality inspection of (1) incoming materials, (2) work in process and (3) final inspection.	Continuous quality monitoring for planning, development, production, marketing and final product usage by consumers.
Standards	Acceptable quality level. Focus on production.	Zero defects. Focus on quality.
Organization	Adversarial relationship between quality control and manufacturing. Quality control is policeman.	Team approach (design, engineering, manufacturing, marketing, sales, etc.).
Quality control	Final inspection to discover defects.	Process control for prevention of defects.

To compete more effectively many food companies are changing the way they manage (assure) quality. There is a desire to develop a total system approach to quality; to develop a quality mentality; to develop a quality culture with all employees having a commitment to quality.

Table 2. Unsuccessful and successful approaches to quality assurance

Unsuccessful	Successful
Downgrade customer's viewpoint.	Emphasize real rather than imagined customer expectations.
Make high quality synonymous with tight tolerance.	Establish what real customer needs are through market research.
Tie quality objectives to manufacturing flow.	Use customer-based quality performance measures.
Express quality objectives as acceptable number of defects.	Formulate quality control systems for all company functions, not just manufacturing.
Formalize quality control systems from the manufacturing function.	

Quality and Profitability

Frank Perdue, the country's best known purveyor of chicken products has said, "People will go out of their way to buy a superior product, and you can charge them a toll for the trek." It is well established that products with superior relative perceived quality images command premium retail prices.

It is the consumer's judgment that is important, not the food company's, and quality perceptions are not absolute, but relative to the competition. A business that has products which are superior in relative perceived quality and which spends more on effective advertising, commands prices that are higher. In addition, advertising influences perceived product quality. If a processor has a high quality product and is able to advertise that product effectively to develop a large market share, the result will be a much higher return on investment than the competition is able to command.

Demographic and Social Changes

Population shifts and changes in life-styles are having important effects on eating habits and on what is considered to be good quality. There are four well established trends that are impacting food quality concerns (Bomben and Borgeson, 1986):

1) The postwar, "baby boom" generation accounts for one-third of the present U.S. population. By the end of this decade this generation and their offspring will impact 50% of all consumer spending. Once considered to be the junk-food generation, they have in their ranks the most vocal proponents and devotees of fresh, natural, healthy foods.

2) By 1990 it is estimated that 75% of all married women 25 to 44 years of age will be working outside the home. About 65% of all children under 6 will have working mothers. The implication of this trend for high-quality convenient (not just quick) foods is profound.

3) America is aging; more than 20% of the U.S. population is over 55 years of age. This population has been called a hidden market and one of the best-kept secrets in the food industry. This group has the most disposable income of any population group. Their more than $500 billion aggregate income almost equals that of the baby boom generation. The older generation has special dietary needs and quality concerns.

4) Another important demographic trend is the relative growth of ethnic populations. In the present decade the white population of the U.S. will grow about 8% whereas the nonwhite population will grow more than 20%. Probably the ethnic group that is having the most effect on American eating habits and food company planning is the Hispanics. They not only represent a significant market, but their foods continue to gain in popularity and importance. Hispanics generally demand higher quality and spend a greater proportion of their income on food.

Americans are changing the kinds of food that they eat. Some of the more obvious trends are the desires for foods that are fresher, more convenient, and all natural. All one has to do is to visit a newer supermarket to gain an appreciation of the important changes that are occurring in food buying habits. The periphery of the stores is increasing in size and complexity while the core area is shrinking. There are large, well-stocked fresh produce, deli, bakery, fresh fish, and meat areas. You can find salad bars and take-home hot meals in these new super stores. Meanwhile, the amount of shelf space in the center of these stores (the dry goods) is decreasing. It is a real challenge for food processors to get into the peripheral areas, which has traditionally been the domain of unbranded foods.

The 24-hour convenience store is becoming a popular place to get a salad or a snack on the run. Stopping off at McDonald's for a salad is becoming increasingly popular. In today's hectic world it is not uncommon to stop off on the commute to grab breakfast or something for dinner at home. In the majority of families all adults work, and convenience is crucial. A recent survey indicated that meal preparations occupy only slightly more than one hour per day for the average U.S. family.

There is a sizeable proportion of the population that seldom sit for a formal meal at home. This group includes the grazers who get much of their food intake more or less on the run.

Consumer Concerns

The latest Food Marketing Institute's survey (Anonymous, 1988) of consumer attitudes showed that the top three quality concerns are taste, product safety, and nutrition. Let's consider each of these in some detail.

Sensory quality
Taste and aroma, are the most important factors in the decision of which foods to buy; 88% of shoppers interviewed by the Food Marketing Institute said that taste is very important. No matter what other attributes a product has, unless it meets consumer's requirements for taste and aroma it will fail. The product attributes which consumers said correlate most closely with good taste are: rich/full flavor, tastes natural, tastes fresh, has a good aroma, and looks appetizing.

A major problem for food companies is to determine what the consumer considers to be the best flavor. There are large differences in individual tastes, and there are significant regional differences in food preferences. Foods that have the greatest appeal to most people in the

North Central and Northeastern United States are quite different than those which are desired by the population of the Southwest with its very strong Hispanic influences. The most successful national food company will be the one that best addresses these regional preferences. This may require regional manufacturing and sales strategies.

Product safety
Safety ranked second in the factors that consumers say influence their decision of which products to buy. Safety is akin to motherhood and whether buying practices parallel concerns is not known. The public is concerned about pesticide residues and microbial contamination. When consumers have a choice, they will likely choose those products which they feel have less chance of having perceived harmful constituents.

When asked how concerned they are about certain items, the consumers felt that the most serious hazards, in order of importance, were: 1) pesticide residues, 2) antibiotics and hormones in meat, 3) fat, 4) cholesterol, and 5) salt (Anonymous, 1988).

There can be no compromise on product safety. Practically, the most important safety issue is microbiological because of potentially acute dangers. The deaths from *Listeria* contamination of dairy products and the publicity regarding *Salmonella* contamination of poultry have raised public awareness of microbiological contamination. The well-publicized Proposition 65 and impending similar legislation in other states, problems with temik in watermelons, alar in apples, heptachlor in dairy products, and ethylene dibromide in cereals have heightened the public concern about pesticide residues (Carnevale, 1988). Crop producers and food processors generally agree on the need to substantially reduce agriculture's reliance on chemicals.

The real hazards of the residues presently found in foods is an open question; some scientists believe that there is a greater hazard from naturally occurring toxicants (Ames et al., 1987). Although public perceptions of the food safety problem do not necessarily square with the facts, it is clear that assuring food products are safe is a crucial quality issue for food processors.

Safety is an issue which must be used carefully in advertising since misuse will exacerbate the misconceptions already held by the public. Consumer protection groups have recommended that an ideal solution to the problems posed by pesticide residues would include: 1)

"Organic foods should be made available in supermarkets so that the consumer has a right to choose between different kinds of produce" (supposedly safe and unsafe). Organic food advocates cannot agree on what is organic. 2) "All produce should be labeled to identify where the food was grown and what pesticide residues it contains." This probably implies that produce of foreign origin has greater pesticide problems than that grown domestically. This issue has become a rallying point for marketing groups in the U.S. trying to curtail importation of produce. Can you imagine what would happen if all food had to be assayed for pesticide residue and then clearly marked as to the residue it contains? The result would be an implied safety problem. 3) "Agricultural production methods should be modified to reduce reliance on pesticide residues (foods should be grown without chemicals)." It is difficult to disagree with this recommendation. However, our plentiful supply of healthful food is dependent to a considerable extent on the use of agricultural chemicals. Food safety is a crucial issue that will continue to be debated and will almost certainly impact consumer buying habits.

Nutrition
The nutritional factors which most concern consumers in order of importance are: 1) fat content, 2) salt content, 3) cholesterol levels, 4) vitamin/mineral content, 5) sugar content, 6) preservatives, 7) freshness, 8) calories, and 9) chemical additives (Anonymous, 1988).

When consumers were asked "How do you react to the nutritional information available to you?", the responses, in order of importance were: 1) Select foods to balance my family's diet; 2) Serve nutritonal snacks such as fruits and vegetables; 3) Check for sodium content; 4) Check the label for caloric content; and 5) Check food labels for protein and fat content (Anonymous, 1988).

About 55% of the consumers surveyed felt that the food in supermarkets is wholesome. However, 53% expressed concern that some ingredients added to processed foods may be harmful or unsafe to eat, and 52% say they avoid buying certain foods because they are concerned about whether they are nutritionally safe.

Affluence, changes in life-style, more working women, smaller households, more eating out, and increased availability of highly processed foods have influenced U.S. food consumption patterns. Some researchers believe that changes in eating habits have led to over-consumption of fat, cholesterol, refined carbohydrates, and sodium. The habit of grazing (eating on the run) is believed by some to have had a negative effect on nutrition; 59 to 70% of U.S. children and teenagers have at least one snack per day, and 40 to 65% of adults consume at least one snack per day. The percentage of the food dollar spent on away from home food consumption has increased from 27% in 1960 to more than 33% in 1970 and is expected to exceed 40% by 1990 (Morgan and Goungetas, 1986).

Despite these changes in eating habits, it appears that where and when people eat does not strongly influence their nutritional well being. Rather, good nutrition requires that people learn to eat in a rational way whether snacking, eating at home, or eating away from home (Senauer, 1986). Regardless of where or when persons eat, they need to consume a variety of foods and eat them in the right proportions. In this country, we are blessed with a wide variety of quality foods. Historically, and in many societies of the world today, human food consumption consists of a simple diet usually dominated by a single staple such as rice, wheat, or corn. Only the rich in such societies can affort varied diets. In sharp contrast, contemporary eating patterns in the U.S. reflect the remarkably varied diet that can occur when economic abundance is widespread (Smiciklas-Wright et al., 1986).

Nutritionists generally accept the idea that eating a variety of foods assures the selection of a nutritionally adequate diet. When asked to define such nutritional concepts as adequate diet and balanced diet, the most likely answer will be eat a variety of foods each day. Such a varied diet must not only provide adequate amounts of the essential nutrients but also must guard against excessive consumption of calories and harmful constituents such as fat and cholesterol.

Fiber, beta-carotene, and crucifers have received attention for their effects on cancer and are worthy of a brief discussion.

Plant fiber has been widely touted for its benefits by the food industry; cereal companies have used fiber content as an important marketing tool. Correlation data suggest a general protective effect against certain cancers by fiber rich foods. There are several metabolic effects of fiber which could play a role in the inhibition of carcinogenesis (Kritchevsky, 1987):

1. Fiber increases fecal weight and decreases transit time.

2. Fiber absorbs organic substances, and it is possible that this absorption has an important effect.

3. Fiber dilutes colonic contents, and there are data which suggest that dilution of bile acids may be correlated with reduced incidence of colon cancer.

4. Another metabolic role attributed to fiber is reduction of energy absorption. Fecal energy excretion in humans fed fruits and vegetables is 92% higher than those fed fruit and vegetable juices. This reduction is most consistent in its effects in reducing spontaneous and induced tumors in experimental animals,

A second component of foods that has received some attention and will likely receive increasing attention in the future is beta-carotene. Beta-carotene is a Vitamin-A precursor and is found in certain fruits and vegetables. Most evidence on the beneficial effect of beta-carotene in reduction of cancer is related to the incidence of lung cancer (Colditz, 1987). It has been shown that both for current smokers and ex-smokers the relative risk of lung cancer in those with a high Vitamin A intake was only 60% of those which had a low Vitamin A intake. Obviously, to stop smoking would have more benefit than changing dietary habits. However, the evidence from epidemiological studies suggests that 15-20% of lung cancers may be attributed to low intake of green and yellow vegetables and most likely to the beneficial effects of beta-carotene.

A group of vegetables receiving considerable attention are the crucifers which includes cabbage, cauliflower, broccoli, brussels sprouts, etc. The American Cancer Society and the National Research Council have recommended increased consumption of cruciferous vegetables as a means of decreasing cancer risk. Data in support of this recommendation come from epidemiological studies, studies with experimental animals, mutagenesis studies, and information on the biological effects of components of crucifers. Consumption of crucifers has been associated with a reduced rate of cancer, particularly in the colon and stomach (Birt, 1987). The most extensively studied agents of these vegetables are the isothiocyanates, indoles and flavonoids. In general, the isothiocyanates and indole-3-carbinol appear to inhibit initiation of cancer.

The studies on the flavonoids seem to indicate that they have antipromotional activity for certain cancers.

Biotechnology and Food Quality

In this section, I will relate briefly some of the recent advances in genetic engineering that will contribute to consumer and agribusiness food quality issues.

Sensory quality

Surveys show that the fresh produce item which has the greatest distance between expectation and performance is the tomato. There are many reasons for this gap, but probably a predominant one is that in order to get tomatoes from winter growing areas to the consumer with an acceptable appearance, they are picked green and the ripening initiated with the naturally occurring plant hormone ethylene. If tomatoes are allowed to remain on the vine until they are ripe, they will meet consumer expectations, but generally this is not possible.

A recent genetic engineering development may help improve tomato quality. The pectin-hydrolyzing enzyme, polygalacturonase (PG) is expressed in tomato only during ripening. PG becomes abundant during ripening and has a major role in cell wall degradation and fruit softening. Recently, tomatoes have been transformed to produce an antisense RNA (Sheehy et al., 1988; Smith et al., 1988). The constitutive synthesis of PG antisense RNA results in a substantial reduction in the level of PG mRNA and the enzymatic activity of ripening fruit. An evaluation of several independent transformants demonstrated that the presence of the antisense RNA was accompanied by up to a 90% reduction in PG activity in red ripe fruit. There is evidence that the transformed tomatoes soften at a slower rate, but that the accumulation of the carotenoid pigments (color) was normal. The use of this same strategy in other fruits or vegetables where a delay in softening could improve consumer quality are obvious.

When it became possible to grow plant cells in liquid culture, it was expected that these cultures could be induced to synthesize a wide range of flavor and other components that contribute to food quality. This enthusiasm has been tempered by the realization that routine plant cell culture production of secondary metabolites is fraught with biochemical, genetic and engineering difficulties.

A potentially useful area of biotechnology, which may have an impact on the production of secondary metabolites by plant cells, is biotransformation. This refers to the ability of plant cells to enzymatically convert a readily available and relatively inexpensive precursor into a more valuable final product.

Traditionally, enzymes and food additives had been used by the processing industry to reduce cost and to improve quality. Currently, the majority of food processing enzymes and additives are produced by biological processes. For the food industry, the application and recombinant DNA technology is becoming a significant part of the product development process. Virtually all classes of food processing enzymes and additives are being targeted by recombinant DNA technology. Trividi (1986) has summarized some of the uses of biotechnology in improving food quality. Certain of these applications will be repeated here to indicate the breadth of present use and possible avenues for future development.

L-glutamate and 5'-nucleotides are the best examples of biotechnology at work for flavor enhancement. The flavor-enhancing properties of L-glutamate were discovered at the turn of the century; today, more than 600 million pounds of this amino acid are produced annually, and more than 96% of it is used in the food industry for flavor potentiation. Virtually all of this compound is produced by biotechnology techniques.

The 5'-nucleotides can be produced by separating and purifying yeast RNA and then treating a crude preparation of the purified RNA with the enzyme phosphodiesterase or nuclease. These enzymes hydrolyze RNA into 5'-AMP, 5'-GMP, 5'-UMP which are used as flavor potentiators. Currently, the nuclease enzyme is isolated from *Penicillium citrinum*. Over production of this enzyme using recombinant technology could be very profitable since such an enzyme could be used to produce 5'-nucleotides in yeast extracts.

Esters impart fruity aromas to foods. Certain microbial cultures smell like pineapple, strawberry or muskmelon because of their characteristic esters. Some of these organisms can be genetically manipulated to enhance their production of esters. The dairy industry uses a diacetyl producing microbe to impart buttery flavor to yogurt and other fermented foods.

Pyrazines are heterocyclic nitrogen containing compounds which upon heat treatment produce a roasted, nutty flavor. Certain bacteria accumulate tetramethyl pyrazines in their culture broth. It is possible that *Corynebacterium glutamicum*, which is used for commercial large-scale production of glutamate, can be used to produce pyrazines. It has been proposed that it should be possible to engineer an organism for production of both glutamate and pyrazines, thereby making production of the latter compounds economical.

Terpenes are important in the flavor of many foods. These compounds are generally produced by higher plants; however, several microbes have been isolated which produce terpenes. As an example, *Kluveromyces lactis* produces the terpenes citronellon and linalool. Biotechnology techniques should be useful to boost yields of these compounds.

Several thousand tons of 1-menthol are used in food, cosmetic, toiletry, and detergent formulations. The plant *Mentha piperita* produces L-menthone during its early flower stage, and this is then converted to L-menthol during late bloom. However, the natural conversion is only 40%. The bacterium *Pseudomonas putida* converts L-menthone to L-menthol, and it may be possible to use this organism to biotransfrom L-menthone to obtain higher yields of a more desirable and racemically pure L-menthol.

Certain varieties of citrus fruit turn bitter as soon as their juice is extracted. This bitterness is caused by certain compounds which are produced from non-bitter precursors. *Aspergillus niger* possesses an enzyme which converts the bitter compound of grapefruit, naringin, to its non-bitter form naringenin.

Other compounds used as flavorants are or can be obtained from biological sources. Doel et al. (1980) synthesized a gene which codes for a protein composed of multiple repeating units of the dipeptide found in aspartame. Proteolytic cleavage of this molecule yields the free dipeptide. The extremely sweet peptides, monellin and thaumatin, are produced by West African plants (van der Wel and Loeve, 1972; 1973; Morris et al., 1973). Their cloning and production in bacteria may be useful in reducing calorie intake or in studies of mechanisms of taste perception.

Approximately $160 million worth of enzymes is consumed annually by food and detergent industries. Amylases, glucose oxidase, lipases, cellulases, proteases, glucose isomerase, and invertase are examples (Haas, 1984). These proteins are sometimes isolated from fungi or other microorganisms which grow at relatively slow rates and may produce only low yields of the product. Others such as pepsin, rennin, and papain are obtained from animal or plant tissue. Microbial production of such proteins achieved by the transfer of their genes to bacteria could provide adequate and stable supplies of material of constant purity and activity. The genes for rennin and a-amylases are among those of industrial importance which have already been cloned. Natural calf rennin is used in cheese making and is in short supply. Calf rennin from genetically engineered fungi is potentially cheaper and could significantly accelerate the cheesemaking process.

Biotechnology advances can reduce the cost of producing food processing enzymes in bulk quantities. It is possible that enzymes may soon be customized for each application. The yield and purity of enzymes can be increased, and properties such as pH and temperature optima and resistance to chemicals can be altered.

Product safety
Implicit in the effort to increase product safety is a reduction in the utilization of chemicals for agricultural production. The use of integrated pest management systems to control pests in agriculture has received increasing attention in recent years. The implications for this effort in increasing product safety are very important. One significant factor in the development of integrated pest management has been the use of biological control. The concept of biological control embodies not only the introduction of antagonists into a cropping system, but also the creation of modified organisms designed to perform particular tasks. The manipulation of the environment to favor resident beneficial organism via crop rotation, residue management, and other cultural practices is important.

There has been increasing attention to efforts to modify the rhisophere colonizing bacteria to provide benefits to crop production through the removal of deleterious microorganisms or chemicals from the environment (Davison, 1988). There are examples which show that it is possible to modify the microbial population of the soil to make it suppressive to certain diseases. An example is certain *Pseudomonas* which are antagonistic to the causative organism of the disease Take All. The ability of *Pseudomonas fluorescens* to suppress Take All of wheat comes from the production of an antibiotic identified as a dimer phenazine carboxylic acid.

A unique kind of soil depolluting bacterium has been described by the team of Ananda Chakrabarty (Karns et al., 1984). They were able to evolve a strain of *Pseudomonas cepacia* which is able to degrade 2,4,5-trichlorophenoxyacetic acid (2,4,5-T), a synthetic herbicide previously considered completely recalcitrant to biodegradation. It was demonstrated that application of this strain to soil contaminated with 1,000 PPM 2,4,5-T reduced the content of the herbicide by more than 98% within one week. The strain used evolved in a highly artificial environment and does not compete well with natural soil bacteria; it is rapidly lost from the soil in the absence of 2,4,5-T. These experiments demonstrate the principle that treatment with suitable bacteria may transform heavily polluted soil to its original condition.

Herbicide resistance is often biochemically and genetically well characterized. Therefore, the engineering of herbicide resistance is feasible. A number of genetic engineering companies are working to develop herbicide resistant crops in the belief that this will reduce the cost of production of the crop. Environmentalists generally believe that herbicide resistant crops will encourage the increased use of herbicides and will be environmentally detrimental because it will discourage the utilization of other techniques which can be used to reduce herbicide use. It is possible that resistance to a herbicide such as glyphosate, which is considered to be "environmentally safe," could reduce the use of herbicides which are less acceptable from safety and environmental points of view. A principal question regarding widespread use of glyphosate would be the effects on the soil microflora which are susceptible to this herbicide.

Since there have been several efforts to develop glyphosate resistant crops, I will focus on this herbicide. The group at Calgene isolated a glyphosate resistant form of the enzyme 5-enolpyruvyl shikimate-3-phosphate (EPSP) from the bacterium *Salmonella typhimurium* (Comai et al., 1985). Glyphosate does not bind to this altered enzyme, thereby giving resistance. The group at Monsanto isolated a cimeric gene from *Petunia hybrida* which overproduces the wild-type EPSP synthase (Shah et al., 1986). This overproduction of the enzyme gives resistance to the herbicide. It remains to be seen whether these two technologies will give a commercially acceptable level of resistance to the herbicide. If the herbicide is not detoxified in the plant and is continually retranslocated to the growing point, these strategies may not be effective.

Much of the pesticide used in this country is for control of insects. Two genetic engineering strategies could have a significant impact on the amount of insecticide used. The first of these is to use a toxin from bacteria to control insects. The second is to create virus resistance plants which will reduce pesticide use since a considerable proportion of the insecticide used is to control vectors of virus diseases; the vectors may have limited economic impact except for the spread of the virus diseases.

The gram positive bacterium *Bacillus thuringiensis* produces proteins which are toxic to a variety of insects. The toxin gene has been cloned from *Bacillus* and has been transferred to a number of crop species (Vaeck et al., 1987). Different strains of *B. thuringiensis* differ in the spectra of their insecticidal activity. Many are active against Lepidoptera but others are specific in their toxicity to Diptera and Coleoptera. A beauty of the *B. thuringiensis* insect toxin is that it is highly specific, and there has been no demonstrated toxicity to other organisms. In addition to the transferring of the toxin gene to plants, it has been engineered into bacteria, particularly *Pseudomonas fluroescens*. By introducing the gene into soil dwelling *P. fluorescens*, the hope is to create a bacterial strain for controlling insects that damage plant roots.

Transgenic tobacco and tomato plants that express the protein coat gene of tobacco mosaic virus (TMV) exhibit a degree of resistance to infection by TMV (Beachy et al., 1987). Although the mechanism of this resistance is not completely understood, it bears a striking resemblance to the resistance in plants that are cross protected with attenuated strains of the virus. The extension of this technology to other virus diseases offers great potential for improving the safety, reducing the cost to produce and extending areas of production of crops, particularly vegetables, to areas where virus diseases now prevent their production. This development could contribute to the consumer's ability to buy safer, lower cost, higher quality, locally grown produce.

The food processing industry has a strong need for naturally produced colors. Artificial colors have been under very close scrutiny by the Food and Drug Administration. The use of microbes to produce natural pigments can be considered. *Phaffia rhodozyma*, a yeast, produces the red carotenoid commonly found in marine vertebrates such as lobster, crabs, and shrimp and in trout and salmon. The fungus *Monascus purpureus* has been used for many years to make red rice wine in the Orient. It actually produces a mixture of red, yellow and purple pigmented polyketides. It has been suggested that by growing *Monascus* under varying conditions, as much as six different shades of pigment could be obtained.

Nutrition

The methods now exist to construct genes that code for synthetic proteins which are enriched in essential amino acid content. The insertion of these man-made genes into the genomes of staple crops will have significant nutritional impact, particularly in the third world countries. Jaynes et al. (1986) have synthesized genes to improve the nutritional quality of potatoes. This same strategy should be possible for corn, cereals and legumes, cassava, etc.

Amino acids are used in the fortification of diets. Currently several hundred tons of amino acids are produced annually (Haas, 1984). Much of this material is obtained by extraction of biological residue or hydrolysates of proteins. Microbial synthesis by hyperproducing organisms may provide a viable alternative to these methods.

The current annual market value of vitamins is at least $670 million. Some vitamins, notably niacin, B12, C and D are made in part or in total by microbial processes. A more efficient synthesis of these compounds may be possible through genetic engineering research.

Conclusion

The speed of advances in biotechnology is breathtaking; it appears that this really will be the revolution so loudly proclaimed in recent years. In the upcoming decades, biotechnologists working with the agricultural and food industry probably will develop ways of improving food quality that have not yet been visualized.

References

Ames, B.N., Magaw, R. and Gold, L.S. 1987. Ranking possible carcinogenic hazards. Science 236:271.

Anonymous, 1988. Trends: Consumer Attitudes and the Supermarket. Food Marketing Institute., Washington, D.C.

Beachy, R.N., Abel, P.P., Nelson, R.S., Rogers, S.G. and Fraley, R.T. 1987. In "Transgenic plants that express the coat protein gene of TMV are resistant to infection by TMV. "Molecular Strategies for Crop Protection," p. 205. Alan R. Liss, Inc. New York.

Birt, D.F. 1987. Anticarcinogenic factors in cruciferous vegetables. In Horticulture and Human Health: Contributions of Fruits and Vegetables. p. 160, Prentice-Hall, Englewood Cliffs, N.J.

Bomben, J.L. and Borgeson, N. 1986. Food Processing Technology: Meeting Changing Industry and Consumer Demands. SRI International, Menlo Park, California.

Carnevale, K. 1988. The safety of fresh produce: Much Ado About Something. Presented to the Food Marketing Institute 1988 Convention on May 10 in Chicago, Illinois.

Colditz, G.A. 1987. Beta-carotene and cancer. In "Horticulture and Human Health: Contributions of Fruits and Vegetables." p. 150. Prentice-Hall, Englewood Cliffs, N.J.

Comai, L., Faccioti, D., Hiatt, W.R., Thompson, G., Rose, R.E. and Stalker, D.M. 1985. Expression in plants of a mutant aroA gene from *Salmonella typhimurium* confers tolerance to glyphosate. Nature 317:741.

Davison, J. 1988. Plant Beneficial Bacteria. Biotech. 6:282.

Doel, M.T., Eaton, M., Cook, E.A., Lewis, H., Patel, T. and Carey, N.H. 1980. The expression in *E. coli* of synthetic repeating polymeric genes coding for poly (L-aspartyl-L-Phenyalanine). Nucleic Acid Res. 8:4575.

Haas, M.J. 1984. Methods and Applications of Genetic Engineering. Food Technol. 38:09.

Jaynes, J.M., Yang, M.S., Espinosa, N. and Dodds, J.H. 1986. Plant Protein Improvement by Genetic Engineering: Use of Synthetic Genes. Trends in Biotech. 4:314.

Karns, J.S., Kilbane, J.J., Chatterjee, D.K. and Chakrabarty, A.M. 1984. Microbial degradation of 2,4,5-trichlorophenoxyacetic acid and chlorphenols. In: "Genetic Control of Environmental Pollutants." p. 3. Plenum Press, New York.

Kritchevsky, D. 1987. The effect of plant fiber consumption on carcinogenesis. In "Horticulture and Human Health: Contributions of Fruits and Vegetables." p. 150, Prentice-Hall, Englewood Cliffs, N.J.

Morgan, K.J. and Goungetas, B. 1986. Snacking and eating away from home. In "What is America Eating." p. 91. National Academy Press., Washington, D.C.

Morris, J.A., Martenson, R., Diebler, G. and Cagan, R.H. 1973. Characterization of Monellin, a protein that tastes sweet. J. Biol. Chem. 248:534.

Senauer, B. 1986. Economics and nutrition. In "What is America Eating." p. 46. National Academy Press, Washington, D.C.

Shah, D.M., Horsch, R.B., Klee, H.J., Kishore, G.M., Winter, J.A., Turner, N.E., Hironaka, C.M., Sanders, P.R., Gasser, C.S., Aykent, S., Siegel, N.R., Rogers, S.R. and Fraley, R.T. 1986. Engineering herbicide tolerance in transgenic plants. Science 233:478.

Sheehy, R.E., Kramer, M. and Hiatt, W.R. 1988. Reduction of polygalacturonase activity in tomato fruit by antisense RNA. PNAS In press.

Smiciklas-Wright, H., Krebs-Smith, S.M. and Krebs-Smith, J. 1986. Variety in Foods. In "What is America Eating." p. 126. National Academy Press, Washington, D.C.

Smith, C.J.S., Watson, C.F., Ray, J., Bird, C.R., Morris, P.C., Schuch, W. and Grierson, D. 1988. Antisense RNA inhibition of polygalacturonase gene expression in transgenic tomatoes. Nature 334:724.

Trividi, N.B. 1986. Food biochnologist concentrate on methods for producing improved flavors and colors. In "Biotechnology in Food Processing." p. 150. Noyes, Park Ridge, N.J.

Vaeck, M., Reynaerts, A., Hofte, H., Jansens, S., deBeuckeleer, M., Dean C., Zabeau, M., van Montaqu, M. and Leemans, J. 1987. Transgenic plants protected from insect attack. Nature 328:33.

van der Wel, H. and Loeve, K. 1972. Characterization of thaumatin I and II, the sweet tasting proteins from *Thaumatococcus danielli*. Eur. J. Biochem. 31:221.

van der Wel, H. and Loeve, K. 1973. Characterization of the sweet tasting protein from *Dicoreophylliem cumminsii* (Stapf) Diets. FEBS Lett. 29:181.

Making Technology Transfer Work

J. G. Ling
Ling Technologies, Inc.
932 Hungerford Drive, Suite #14
Rockville, MD 20850, USA

Increasing competition in the international marketplace has made the U.S. more concerned about making better use of its technological resources. One way to do this is to achieve greater cooperation and synergism in R&D between universities, federal laboratories, and industry. Technology transfer involves the commercialization, by industry, of technologies developed in the other two performing sectors.

The federal government has instituted new laws and policies to encourage technology transfer from the laboratories to private industry. The process, however, is still sluggish because of entrenched attitudes, unfamiliarity with opportunities and procedures, and bureaucratic inertia on both sides. Ultimately, technology transfer is a people-to-people process that requires personal interaction and commitment by the participants. A technology transfer broker or facilitator can be helpful in bringing the two parties together and in expediting the flow of paperwork through the corporate and federal bureaucracies.

Introduction

A great deal of U.S. government and private-sector attention is being directed today at the issue of international industrial competitiveness. In particular, there is concern that the U.S. is rapidly losing the position of technological dominance that it held twenty years ago. In terms of sheer numbers, however, the U.S. science and technology base still commands a leading position in the world. The U.S. continues to spend more money on research and development than Japan and Western Europe combined. This spending translates into more researchers, universities, and laboratories than these other nations have. There is a general feeling, however, that the country is not making the most effective use of those resources toward improving its industrial position; and efforts have been made by the government in the past few years to encourage, facilitate, and expedite the transfer of technology from universities and federal laboratories to industry. Stated another way, the government is anxious to have more commercial products evolve from the large national research base. The university-industry link is fairly well established by this time, and this paper focuses on the link between federal laboratories and industry which is still in a developmental stage.

Federal Laboratories as a Resource

The U.S. government spends close to $20 billion a year to operate over 700 research entities. Of these, 146 have 100 or more professionals on their staffs, and within that group there are 51 that have more than 500 professionals. The laboratories of primary interest to those involved in food quality and biotechnology are operated by the Agricultural Research Service, National Institutes of Health, Food and Drug Administration, and the U.S. Army. These laboratories have a wide range of scientific activities with special emphasis on basic and applied "high risk" research which U.S. industry normally would not undertake.

A number of recent actions taken by Congress and the Administration have encouraged the transfer of federal laboratory technology to U.S. firms. Federal laboratories can now negotiate cooperative research and development agreements with members of the private sector, providing added resources to the laboratory and giving private-sector participants exclusive rights to commercialize the results of such research.

There are several ways in which cooperative industry-federal laboratory R&D can take place. The general types of cooperation are:

User facilities
In this type of cooperation, industry pays the laboratory a standard fee for using a piece of equipment or a facility at the laboratory (e.g., the synchrotron light source at Brookhaven National Laboratory).

Contract research
The laboratory performs certain research services (e.g., product testing) that are not available in the private sector.

Team approach
The company and federal laboratory work together on a project, each contributing part of the necessary intellect, manpower, and facilities

(e.g., laboratory does basic research, then teams with company to do prototype development).

Equipment contribution
The company provides a needed piece of equipment to the federal laboratory as a loan or gift in order to expedite research of mutual interest. This contribution can be very helpful to the laboratory because the company can usually move faster than the government procurement office.

Personnel contribution
The company provides research manpower to the federal laboratory in the form of a guest researcher, postdoctoral researcher, or laboratory technician. Many laboratories welcome such help because of government manpower ceilings.

Through these mechanisms, private companies can take advantage of the equipment, expertise, and ideas resident in the federal laboratories. They can also, in many cases, get exclusive rights to any government patents resulting from the cooperation *before* the cooperative effort begins.

On the laboratory side, the inventors are now guaranteed a minimum share of royalties on their patents. So there is a much greater incentive for commercialization than there was in the past. Also, the laboratories will receive the balance of the royalties which, in the past, went into the general receipts of the U.S. Treasury.

One way of looking at this new situation is that industries can now negotiate agreements with federal laboratories that are very much like the agreements between industries and universities. They include protection of proprietary information but allow publication of research findings. The advantage of dealing with the government is that, unlike a university, it is able to contribute substantial amounts of its own resources as part of the bargain.

Tapping Laboratory Technology

Opportunities for taking advantage of federal laboratory technologies are still underutilized. Some of the reasons for this underutilization are:

- Lack of awareness on the part of the U.S. private sector that marketable technologies exist in federal laboratories.
- Lack of time, know-how, or contacts on the part of laboratory staffs to stimulate commercial interest.
- "Not invented here" syndrome exhibited by U.S. private industry.
- Reluctance of the U.S. private sector to get involved with the federal bureaucracy.

The government has done a lot to ease the way for laboratory cooperation with industry. However, aside from creating a good internal environment for technology transfer, the government's role is necessarily passive. The various federal agencies can establish standard procedures for handling cooperative research and development agreements with industry. They can also provide data bases describing research results or work in progress. However, they cannot target specific companies and actively solicit their cooperation. They do put on "show and tell" sessions to describe cooperative opportunities to a group of companies, but these have not proved very productive because they do not provide the one-on-one hard-sell approach that seems needed.

If technology transfer from federal laboratories is to work, private industry must show more interest and play a more active role. At the present time the field is relatively uncrowded, and an enterprising company has an opportunity to participate in the most promising technologies in the laboratories. To do so, however, requires getting to know the researchers involved. Ultimately, technology transfer is a people-to-people activity that cannot be replaced by publications, data bases, or displays. Company researchers and research managers must establish rapport with their counterparts in the federal laboratory as a precondition to any collaborative effort. Experience has shown that only by visiting the laboratories and talking to the researchers can someone in industry find out what is really going on. Many of the exciting opportunities are unpublished and can only be discovered through personal discussion.

Two examples illustrate how the cooperative research and development process can work. In one case, the federal laboratory is developing monoclonal antibodies for a variety of plant diseases. In order to commercialize this product, they have teamed up with a small company that performs diagnostic services and sells diagnostic kits. The company tests the use of antibodies under field conditions and feeds back information to the laboratory. Together they come up with a product that meets industry needs and can be marketed by the company. In another case, a small company doing research on transgenic animals is developing new techniques for genetically modifying animal embryos to improve food quality. In teaming up with a federal laboratory it gains the expertise to test these techniques on animals. Success in this cooperative effort can lead to the emergence of greatly improved food animals.

Thus far, small and medium-sized companies have been more inclined to participate with the federal laboratories than large ones. Large corporations move more slowly, and their own bureaucratic procedures tend to inhibit the introduction of technologies from outside sources. The use of technology brokers to help companies, both large and small, find cooperative opportunities in the laboratories can be very useful. A competent broker will already have a good rapport with the laboratory of interest and can identify outstanding researchers. This kind of help can greatly reduce the search time for a company, and a good broker can also help move cooperative agreements through the corporate and governmental approval processes.

The real opportunities in technology transfer lie in finding laboratory projects that are in the pre-prototype or early prototype stage and are perceived by a particular company to have good market potential in its line of business. Then through a formal cooperative research and development arrangement the company can guide further development and ensure itself an exclusive position in the marketplace for the final product.

This type of technology transfer is labor intensive and involves a lot of personal interaction between the facilitator, company, and laboratory. In the end it boils down to a people-to-people activity involving goodwill, trust, and mutual respect among all the participants. It also takes time and patience.

This new era in technology transfer from federal laboratories offers exciting possibilities for companies seeking new ideas and ways to leverage their own investment in research and development, and rewards await those who move quickly to stake out the most promising areas. The laboratories will gain as well through the input of new ideas and additional resources.

Biotechnology: Regulatory Considerations

Kenneth A. Gilles
Assistant Secretary for Marketing and Inspection Services
U.S. Department of Agriculture, USA

I would like to address the topic of biotechnology from a perspective of discussing how USDA will regulate the products of biotechnology, particularly meat and poultry.

Certainly, very few issues have aroused greater interest and have been as misunderstood as biotechnology. On one hand, scientists point to the fortuitous development of one of the "greatest scientific breakthroughs of our time—the discovery of the structure and importance of DNA," the makeup of that remarkable double helix string of life that carries the genetic messengers with instructions for all living cells. Now, through biotechnology methods, we can identify good DNA genetic messages in one species . . . and shift them to another species.

It couldn't happen naturally. Different species usually are sexually incompatible and don't breed. Borrowing good traits speeds up the ability to improve our quality of life. I believe we all share the confidence that biotechnology offers the creative potential to solve some of the more difficult problems facing the agricultural community today.

On the other hand, laymen read articles such as appeared in *USA Today*:

> "Picture this farm: super dairy cows getting hormone shots to produce more milk. Soybeans killing the insects and weeds that prey on them. Steers, injected with a human growth hormone, as big as elephants."

Well, I certainly haven't seen any cattle as large as elephants walking around USDA, but this is the type of picture that may come to the public's mind when the words "biotechnology" or "genetic engineering" are mentioned.

I find this unfortunate, because this sensationalism obscures the true benefits that such technology has brought, and will bring. For instance, in the area of animal health, genetically engineered animal vaccines, which are safer and more effective than traditional vaccines, have saved livestock producers millions of dollars each year. And we are getting closer to the day when animals can be custom-designed to resist disease. USDA scientists recently developed a new strain of chicken that appears to be resistant to avian leukosis, a disease that causes considerable losses to the poultry industry each year.

Not only does biotechnology offer us the ability to improve animal health, it also provides us with tools to make animal products healthier for consumers. Daily bovine somatropin injections, for instance, can produce pigs with less fat, a worthwhile goal in light of recent efforts to reduce fat in the American diet. In fact, the National Research Council, in its 1988 publication entitled *Designing Foods—Animal Product Options in the Marketplace*, encouraged the development of new technologies such as genetic engineering that can produce animal products more in line with U.S. dietary goals.

It behooves all of us to enhance biotechnology's public image by dispelling the myths and helping the public to understand what this new technology will really mean to the future of agriculture and mankind.

Coordination of Biotechnology Activities in USDA

At USDA, we allay consumer concerns about the safety of biotechnology through careful coordination of all research and regulatory activities.

There are seven agencies within USDA currently involved in biotechnology. Three are involved in biotechnology research—the Agricultural Research Service, the Cooperative State Research Service, and the Forest Service. Three are regulatory agencies—the Animal and Plant Health Inspection Service, the Agricultural Marketing Service, and the Food Safety and Inspection Service. The seventh agency is the Economic Research Service, which evaluates the socioeconomic impacts of new agricultural technologies.

To help guide the activities of these agencies, we have established the Committee on Biotechnology in Agriculture, co-chaired by Dr. Orville Bentley, Assistant Secretary of Agriculture for Science and

Education, who has responsibility for research, and myself, who has the responsibility for regulation. The committee meets to discuss and resolve scientific and policy issues. It functions as both a policy body and a bridge between research and regulation within the USDA. You see there is a clear separation of authority for research and regulation in the USDA.

We also have established an Office of Agricultural Biotechnology. This staff unit has responsibility for coordinating USDA procedures pertaining to biotechnology, including laboratory and field research.

A third group is the newly-formed Agricultural Biotechnology Research Advisory Committee, which includes representatives from government, industry and academia. This advisory committee helps to promote proper scientific procedures used for USDA-funded biotechnology research.

In addition to these committees, each individual agency may have additional groups established to meet their specific needs.

USDA Philosophy on Biotechnology

Within USDA our goal is to foster a climate that encourages innovation, while at the same time ensuring that agricultural research is carried out safely and that the products of biotechnology are regulated based on firm scientific principles.

To ensure consistency as we go about this challenging task, USDA, along with other federal agencies that are members of the Biological Science Coordinating Committee (BSCC) of the Office of Science and Technology Policy (OSTP) in the White House, stated its policy for regulating biotechnology in December 1984 and June 1986, as part of the Federal Register Notice entitled "Coordinated Framework for Regulation of Biotechnology." Dr. Bentley and I, representing the USDA, participate in the BSCC, the Committee of Life Science of the OSTP, and the International Biotechnology Committee of the Department of State.

USDA policy is based on three primary elements. First, we believe that the product and the risk, rather than the technology, should be regulated. Genetic exchange has existed for thousands of years. Initially, it occurred spontaneously! Later, breeders and farmers introduced desirable genes into plants and animals through selective breeding. The technology developed over the past 15 years or so has allowed us to accomplish these goals faster, and with greater precision. Consequently, we do not believe that biotechnology products should be treated any differently from traditional products. In fact, one might argue that the greater precision offered by this new technology may *add* a degree of product safety.

We will pursue our regulatory responsibilities with respect to these new products with the same rigorous review we apply to more conventional products. We will use existing regulations, including the plant pest regulations published in July 1987 specifically for pests modified by genetic engineering.

Second, we do not believe that products developed through biotechnology will differ markedly from conventional products. Descriptions of cows as big as elephants make good news copy, but in reality, we do not anticipate major changes in the products of biotechnology.

Third, we believe that the existing statutes are adequate for regulating products of agricultural biotechnology. However, we will reevaluate periodically the existing regulatory framework to determine if there is a need for addition, modification, or reduction of regulatory measures.

Regulation of Meat and Poultry

I'd like to elaborate on our philosophy in a little more detail by discussing how we intend to regulate meat and poultry products produced through, or affected by, biotechnology techniques.

Biotechnology most likely will affect food animals in two ways, by the introduction of exogenous, bioengineered substances or by the transfer of new genetic material into the genetic makeup of an animal.

In the first case—the exogenous introduction of bioengineered substances—a number of safeguards exist to ensure adequate regulation. Initially, the appropriate federal agency must approve the use of any of these substances. In the case of biologics such as vaccines, USDA would evaluate the substance for safety, purity, potency, and effectiveness. Drugs, on the other hand, would be evaluated by the Food and Drug Administration, and pesticides, by the Environmental Protection Agency.

In addition to a careful review of the substance by the appropriate regulatory agency, we have regulations governing the use of animals administered these substances for food. No animal used in any research investigation involving an experimental biological product, drug, or chemical shall be eligible for slaughter for use as food unless it can be demonstrated that the animal is not adulterated.

In the second case—the transfer of new genetic material into the genetic makeup of an animal—our regulatory response will depend on the type and magnitude of the genetic change induced. In general, we do not anticipate that bioengineered animals will differ substantially in appearance, behavior or general health from currently inspected animals. In most cases, there may be an alteration in one, two, or a few genes that may change the animal's growth rate or disease resistance without significantly altering its basic physical appearance, general health, or behavior. These animals would be inspected like their more traditional counterparts. I should emphasize that no such aniamls have yet come to market.

A more complicated situation arises in the case of food animals whose cells or tissues contain different genetic information. For instance, in 1984 laboratories in the United Kingdom and West Germany reported the occurrence of transgenic animals, real-life chimeras, between sheep and goat. These animals were produced by embryo fusion, and the adult animals had the horns of a goat and fleece of a sheep.

The likelihood of an animal such as this being produced for a reason other than pure research is highly remote, but it does present some philosophical questions. For instance, what do you call such an animal? Since USDA must ensure that meat and poultry products are accurately labeled, this would be a consideration. Another question is whether the animal is subject to mandatory federal inspection. Under the Federal Meat Inspection Act and the Poultry Products Inspection Act, all cattle, sheep, swine, goats, equines, poultry, and food products prepared from these animals must be inspected, and at government expense. The sheep-goat hybrid would certainly fall into this category. But what about a hybrid between a cow, which must be inspected, and a buffalo, which is subject only to voluntary inspection?

As a side note, this situation actually did arise, not as a result of genetic engineering, but of traditional crossbreeding. We were presented with two hybrid crosses of cattle and buffalo. One hybrid, called "Cattalo," resembled a buffalo. The other, called "Beefalo," more closely resembled a cow. Even though both are hybrid crosses of cattle and buffalo, we decided that only the beefalo was covered under the Federal Meat Inspection Act since it resembled cattle. The other hybrid was considered exempt from mandatory inspection.

Closing

Perhaps these scenarios will never come to pass, largely because the economic incentive to produce these novel animals just isn't there yet. I believe we will see the largest impact in the use of biotechnology to improve the health of food animals, and to alter the nutritional content of meat and poultry.

It is important, however, to imagine the possibilities so we can anticipate where technology may take us. This is the only way we will meet our goal of fostering innovation while at the same time ensuring the products of biotechnology are properly regulated.

As we enter the last decade of this century, agriculture's job is tougher. We are challenged to provide the highest quality of food and fiber at lower costs. From abroad comes the challenge of competition for international markets. Here at home we are challenged to protect plant and animal health, while providing maximum protection for human health and the environment. How to do that? The answer lies in research, development, adaptation, and education. And we'll succeed!

Biology: Now Is the Time

Ralph W.F. Hardy
*President, Boyce Thompson Institute
for Plant Research at Cornell
Ithaca, NY 14850, USA
and
Deputy Chairman, BioTechnica International, Inc.
85 Bolton Street
Cambridge, MA 02140, USA*

It is a pleasure to participate in this timely International Symposium on Biotechnology and Food Quality organized by UM-USDA-DuPont. As we all know, we are living in an increasingly competitive world. The most successful universities, government agencies, and industries will, in a synergistic manner, partner their capabilities to strengthen their competitiveness without forfeiting their unique major individual roles. This meeting is an example of an early stage of such a partnership in the area of biotechnology and food quality. I commend the organizers present this evening—Shain-dow Kung, Donald Bills, and Ralph Quatrano—for bringing together their respective host organizations.

World Change and Biology

Our world is at a stage of dynamic, exciting, and challenging change. A recent report, published by DuPont, projected some of the expected changes from 1990-2020, three decades at the end of one century and the beginning of the next. World population was projected to grow 42 percent while U.S. population will remain almost constant with an increase of only 9 percent. The growth market for food, at least in numbers of consumers, will be outside the U.S. Cost of staple foods like milk and bread are expected to increase over 400 percent while for comparison purposes the cost of an average new home in the U.S. is projected to become $2 to $3 million, or 1,400 percent greater than in 1990. Sales of the largest corporation are projected to increase by 2,600 percent, undoubtedly representing horizontal and probably vertical consolidation. Just today Philip Morris offered $11 billion to buy Kraft and create the largest consumer products company. Where will this largest company be headquartered in 2020—U.S.A.? Japan? Another country? The average life span of man is projected to increase by 21 percent and that of woman by only 15 percent; the long-standing sexual discrimination in aging against men will finally be overcome with an average life span of 91 for each of men and women in 2020. In the political area, a woman president is projected. Executives and professionals like the attendees at this meeting may be pleased to know that their work week will increase by 3 percent—to 60 hours—while laborers will decrease theirs by 18 percent—to 32 hours. These projections may or may not prove to be true.

However, there is one projection about which I am most confident, and this projection will have major impact not only on the subject matter of this symposium but on almost all aspects of the quality of human life. This projection is that biology is emerging as the dominant science and during this thirty-year period will not only be the dominant science but will also drive the innovative process as engineering, chemistry, and physics did for most of the 20th century.

I will use an example with which we are all only too familiar to document the emerging power of biology. In 1980, less than a decade ago, a new disease with the unusual characteristics of location in certain east and west coast cities, prevalence in males with a certain life style, and destructive effects on the immune system was recognized. Within two to three years the causative agent was identified in U.S. and French laboratories as the HIV virus, and by 1985 the molecular structure of the 10,000 base HIV genome was characterized by a team of DuPont and NIH scientists. I had the good fortune to head life science at DuPont when several of these scientists joined the exciting DuPont life science expansion of the late 1970s and early 1980s. Several companies including DuPont developed effective diagnostics by 1985 and the safety of the blood supply was assured. By 1987 the first AIDS vaccine received FDA approval for clinical trials. This significant vaccine was produced by a modified insect virus whose biology was elucidated by agricultural scientists. A decade from now, I believe that our capabilities in biology will have produced products and processes with such efficacy that AIDS will then be viewed in the same way that we now view smallpox and polio. Without the biology base of the 1940s to 1980s and the current momentum in biology, the above accomplishments in less than a decade would not have been made and future expectations would not be possible.

Era of Biology

We are living in the era of biology. This biological era will provide innovative products and processes for health, agriculture, food, environmental clean-up, energy, mining, chemicals, and maybe even information handling. Biology is changing from an experimental-observational science to a theoretical-predictive science. The fundamental molecular laws of biology are being discovered with universality across organisms. These fundamental laws include:

- DNA/RNA as genetic material
- DNA as self replicating
- Linear DNA —> linear RNA —> linear protein with the theoretical ability to write the linear molecular structure of protein or RNA from DNA
- Linear protein —> three-dimensional protein (law(s) not yet known)
- Three-dimensional protein —> biological activity (laws not yet known).

Molecular biologists are proposing to map and sequence human and hopefully other genomes. The sequencing of a plant chloroplast genome with about 150 kilobases was reported in 1986. Arabidopsis is a higher plant of the mustard family and is of great interest for mapping and sequencing because it has only 70,000 kilobases while most other higher plants have 3 to 6 million kilobases, similar to the 3 million kilobases in the human genome. From mapping and sequencing activities will come an explosive expansion in the biological data base, thereby facilitating the theoretical and predictive phase of biology.

We have already seen an upward discontinuity in our understanding of immunology and neuroscience and expect to see such in plant, animal, and food science. The latter are trailing because of the inadequate base of public funding of the agricultural and food sciences during the last several decades. Last fall in congressional testimony, I encouraged a competitive grant program for new initiatives research in agriculture and food with funding to be increased in a step-wise fashion over a 5-year period to $500 million annually.

Many of the tools of the biological tool kit are in hand. This is especially the case for nucleic acids and somewhat for proteins. We can identify genes—possibly a single molecule—via amplification or enhancement technology. In 1985 I was involved in the establishment of a company to commercialize DNA probe diagnostics for the microbes that are involved in periodontal disease. This disease is the major cause of tooth loss in people over 35 in the U.S. and a disease that costs $4,000 million annually for treatment. DNA probe technology has become impressively sensitive in detection of possibly as little as only one DNA molecule. Note the forensic applications reported in the popular press. We now have tools to begin to map and sequence large genomes. We can synthesize genes in the laboratory and we can even design genes. We can transfer genes to many plants, animals, and microbes; and these genes function in their new homes. We can mark genes to follow their presence and functionality. We are determining gene function in disease. Only one of the eight genes of cauliflower mosaic virus produces all the symptoms of this disease; somewhat similar observations have just been made with one of the HIV genes. We can modify gene function with different promoters, gene amplification, synthetic chemicals, and most recently, anti-sense genes where a laboratory example demonstrated a potential prolonged shelf life of tomatoes.

A similar but less potent list of biological tools for proteins could be listed. The relative status of the protein to nucleic acid tool kit is documented by a statistic. The amino acid sequence of about 5,000 proteins is known; however, most were inferred from DNA sequences rather than obtained from direct amino acid sequence data. Of relevance to agriculture and food is the additional fact that most of these sequences are for human proteins; few are for those from plants. The above momentum and tools document that biology's time is now, and we must increase support for agriculture and food research so that these areas can benefit from the biology era as is health care already.

Products and Processes of the Era of Biology for Food and Food Quality

The era of biology will provide high value-in-use diagnostics, microbial products, microbes, seeds, and animals for food and food quality.

We are early in the development and commercialization stage for these products and processes.

The technology is in hand for the development of any diagnostic product for which there is a justifiable need. This technology includes immuno-based assays and DNA-based assays including RFLPs and probes. There are markers like GUS and luciferase to follow genes and gene function. These diagnostics will be used to measure or detect quality, impurities, composition, disease, pests, resistance, and proprietariness, and to facilitate or guide processes such as plant and animal breeding. These diagnostics will become user friendly allowing the consultant, production farmer, processor, or consumer direct and rapid measurement.

Microbial products such as biopesticides, animal hormones, and synthetic vaccines will improve the quality of food. Biopesticides will achieve this by reducing use of synthetic chemical pesticides on plants. The first synthetic vaccines for animals are already commercial while animal hormones have produced outstanding results in research and development trials in improving "healthfulness" of meat.

Microbes are being developed as biopesticides, protective agents against physicochemical stress, agents for improved nitrogen input and yield, and as post-harvest treatment products for plant agriculture. Although *Bacillus thuringiensis* and its toxin have received much emphasis as a biological control agent for insect pests, there are viral and fungal agents that may have even greater potential. Improved rhizobia are expected to increase substantially the yield of leguminous plants. Let me note a recent unexpected discovery in this area. It illustrates unused or unrecognized diversity in living organisms, especially microbes. At Boyce Thompson Institute we have discovered rhizobia that are photosynthetic as well as N_2-fixing, thereby having the potential for energy self-sufficiency in N_2 fixation.

Seeds will become the major product from the era of biology for plant agriculture. Agronomic characteristics will be improved to increase productivity in the near term. These agronomic characteristics will overcome biological and physicochemical stress and increase yield. In addition, these seeds will, in the longer term, produce higher value-in-use crops with improved processing characteristics and/or enhanced consumer preference characteristics. Examples of improved processing characteristics are increased tomato solids and modified oil, carbohydrate, or protein composition. Enhanced consumer preference characteristics will have the highest economic value. These characteristics include decreased chemical residues, improved taste, flavor, and color, extended shelf life, and improved healthfulness.

In the longer term, animals modified by the era of biology will become important for food and food quality. This technology base is least developed. Clearly, the era of biology will create improved products and processes for existing markets and generate new products and processes for new markets. Products and processes based on chemistry and engineering will be replaced by those based on biology with favorable impact on quality, economics, and also the environment. A variety of other factors are key to development and commercialization of the products and processes of the era of biology.

Other Key Factors

The availability of proprietariness for biological innovations is increasing in the U.S. and probably in the world. The U.S. Patent and Trademark Office (PTO) awarded the first patent for a living organism, a bacteria that degraded oil, in 1980. Four years later the first PTO patent for a sexually propagated plant was awarded, and in 1987 the first PTO patent for an animal was awarded. This increased proprietariness for biological materials is due to a number of factors:

- Technology is molecular, like chemistry.
- Research and development is more costly than selection and traditional breeding.
- High value-in-use is anticipated.
- Need for proprietariness to obtain appropriate return for cost of development.
- National and international competitiveness in biotechnology.

My projection is that biological products will be protected under the PTO, enabling attractive returns for high value-in-use products.

Private industry will be essential for the development and commercialization of the products and processes of the era of biology. The U.S. is fortunate in having about twenty large, established companies and about ten development-stage companies with major commitments to products and processes for the food-quality market. We in the U.S.

have an unequally strong position in commercialization of these products.

We are living in a pro-regulatory environment. Over twenty laws have been enacted since 1950 for federal protection of the environment. Products and processes of a new technology will be more regulated than those of earlier technologies. The products and processes of the era of biology will be more regulated than those of plant and animal breeding. This is scientifically incorrect since the products of the era of biology are less risky because of the directed and defined nature of the new process versus the random and undefined nature of the traditional process. Nevertheless, progress is being made in field testing of genetically engineered microbes and plants. A 1987 workshop meeting on genetically engineered plants and agricultural and environmental risks concluded that for defined crops and changes there is negligible risk (Boyce Thompson Institute. *Regulatory Considerations: Genetically Engineered Plants*. Summary of a Workshop Held at Boyce Thompson Institute for Plant Research at Cornell University, October 19-21, 1987).

In conclusion, we have entered the era of biology. Early products and processes for the food area are being developed and commercialized such as diagnostics, microbial products, microbes, and seeds; and in time, these will include animals. These products and processes will have high value-in-use and improve the quality of food.

GENETIC ENGINEERING
AND FOOD QUALITY

Characterization and Modification of Maize Storage Proteins

Brian A. Larkins
Department of Plant Sciences
University of Arizona
Tucson, AZ 85721, USA

John C. Wallace and Craig R. Lending
Department of Botany and Plant Pathology
Purdue University
West Lafayette, IN 47907, USA

Gad Galili
Department of Plant Genetics
Weizmann Institute
Rehovot, Israel

Evelynn E. Kawata
Department of Biology
Yale University
New Haven, CT, USA

Introduction

With the development of recombinant DNA and genetic engineering techniques, there has been growing interest in the application of these procedures to the improvement of crop plants. The development of high-yielding, low cost-input crop plants, which has been done traditionally through plant breeding, could be accelerated through the utilization of the techniques of modern molecular biology. While many potential targets for this research have been identified, the improvement of the nutritional value of seed storage proteins is one which can be immediately approached.

Storage proteins account for the majority of protein in most seeds. In general, these proteins are deficient in one or more amino acids that are required for human and livestock nutrition. The major storage proteins in legumes, 11S and 7S globulins, are deficient in methionine and cysteine, while the storage proteins of cereals, prolamines, are deficient in lysine and tryptophan. With current technology, the deficiencies of these amino acids are compensated by the addition of amino acids or blending of feed rations. These costly procedures could be eliminated through genetic engineering of genes encoding seed storage proteins.

In recent years, genes encoding the seed storage proteins of most major crops have been isolated and sequenced, and the cellular mechanisms responsible for the synthesis and deposition of these proteins into protein bodies have been have been characterized (Shotwell and Larkins, 1989). With the development of gene vectors and plant transformation systems, it has also been possible to create transgenic plants that express these genes. Much remains to be learned about the regulation of transcriptional and posttranscriptional mechanisms to enhance the expression of these genes and maximize the amount of protein synthesized. This will undoubtedly be the focus of future research efforts.

Before it is possible to devise an approach for the genetic engineering of seed proteins, it is important to understand the structure and interaction of these proteins within the developing seed. The genes encoding storage proteins must be modified in such a way that the inserted amino acids do not deleteriously affect the normal biological processes associated with the synthesis and deposition of these proteins. It is with these questions in mind that we have been studying the storage proteins of maize seed.

Results and Discussion

Synthesis and structure of zein proteins

Maize seed storage proteins are a group of alcohol-soluble polypeptides that are called zeins (Esen, 1986). Zeins are synthesized in the endosperm of the developing seed and at maturity account for 50% of the total endosperm protein. These proteins are synthesized by polyribosomes bound to the rough endoplasmic reticulum (RER) and aggregate within the lumen of the RER to form organelles called protein bodies (Larkins and Hurkman, 1978). Zeins can be extracted from mature endosperm with 70% ethanol or 55% isopropanol in the presence of a reducing agent such as 2-mercaptoethanol; alternatively, they can be solubilized from protein bodies isolated from developing seeds.

Analysis of the protein body proteins reveals a mixture of polypeptides with Mr of 27,000, 22,000, 19,000, 16,000, 14,000, and 10,000 (Fig. 1). The primary amino acid sequence of these proteins reveals four distinct structural types that we designate alpha, beta, gamma, and delta zeins. Proteins of the alpha-type, Mr of 22,000 and 19,000, usually account for approximately 60% of the total protein, based on Coomassie Blue staining. The beta (Mr 14,000), gamma (Mr 27,000 and Mr 16,000), and delta-zeins (Mr 10,000) account for about 10%, 25%, and 5%, respectively, of the total.

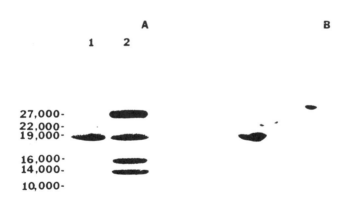

Figure 1. SDS-PAGE separation of maize storage proteins. (A) Proteins extracted from mature endosperm with 70% ethanol alone (lane 1) or 70% ethanol plus 2-mercaptoethanol (lane 2). (B) Separation of protein body proteins from developing endosperm by non-equilibrium isoelectric focusing (left to right) and SDS-PAGE (top to bottom). Proteins in both gels were visualized by Coomassie Blue staining. Apparent mol wts are indicated on the left (from Larkins et al., 1984).

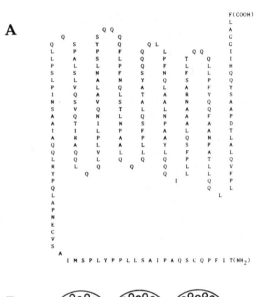

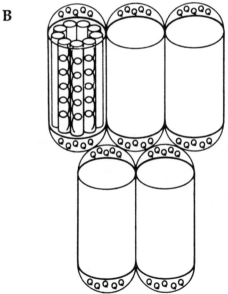

Figure 2. Structure of alpha zein proteins. (A) Primary amino acid sequence of Mr 19,000 alpha zein; following the NH_2-terminus, the repeated peptides are illustrated folding back and forth. (B) Proposed structural model. Repeated peptides form alpha helices that wind up and down and fold the protein into a rod-shaped molecule (from Argos et al., 1982).

These four types of zein proteins are structurally distinct. Alpha zeins typically have an NH_2-terminus of 40 amino acids that precedes a series of eight to ten repeated peptides of approximately 20 amino acids. These repeats, which are tandem, appear to be alpha helices that fold the protein into a rod-shaped structure (Fig. 2) (Argos et al., 1982). The beta-zein, which is rich in sulfur-amino acids, has no repeated peptides and has primarily sheet and turn conformation (Fig. 3) (Pedersen et al., 1986). The gamma zeins have an exceptionally high content of proline. Little is known about the structure of these proteins (Prat et al., 1985), but because the NH_2-terminal half of the Mr 27,000 gamma zein is more than 50% proline (Fig. 4), it would be expected to have a very extended conformation. The delta zein, like the beta zein protein, is rich in methionine and cysteine (Fig. 5); however, at present little is known about its structure (Kirihara et al., 1988).

β Zein
S-Rich Mr 14,000 Protein

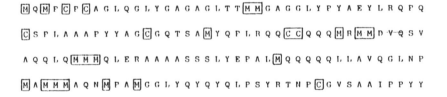

Figure 3. Primary amino acid sequence of Mr 14,000 beta zein. The NH_2-terminus of the protein begins at the upper left and proceeds left to right. Starred positions indicate regions of homology with the Mr 27,000 gamma zein. On the scale in the lower part of the figure the tall bars mark the positions of methionine residues and the short bars the positions of the cysteine residues.

Because the beta, gamma, and delta zeins require the presence of a reducing agent for solubility, they must be cross-linked by disulfide bonds. As yet it is unclear whether cross-linkages form among polypeptides of the same type or between polypeptides of different types.

Localization of Zeins Within Protein Bodies

The different types of zein proteins are localized heterogeneously within protein bodies. Differences in the compositions of protein bodies are suggested by variations in their staining intensity with lead citrate and uranyl acetate (Fig. 6). Many protein bodies have a homogeneous electron-lucent matrix, while others have dark-staining material which occurs at the surfaces and/or as inclusions within the protein body (Fig. 6A).

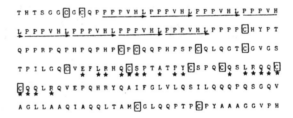

Figure 4. Primary amino acid sequence of Mr 27,000 gamma zein. The NH_2-terminus of the protein begins at the upper left and proceeds left to right. Arrows mark the positions of hexapeptide repeats; starred positions indicate regions of homology with the Mr 14,000 beta zein. On the scale in the lower part of the figure, the short bars mark the positions of proline residues and the tall bars the positions of cysteine residues.

δ Zein

Methionine-Rich Mr 10,000 Protein

T H I P G H L P P V [M] P L G T [M] N P [C] [M]

Q Y [C] [M] [M] Q Q G L A S L [M] A [C] P S L [M] L

Q Q L L A L P L Q T [M] P V [M] [M] P Q [M] [M] T

P N [M] [M] S P L [M] [M] P S [M] [M] S P [M] V L P S

[M] [M] S Q [M] [M] [M] P Q [C] H [C] D A V S Q I [M] L

Q Q Q L P F [M] F N P [M] A [M] T I P P [M] F L

Q Q P F V G A A F

Figure 5. Primary amino acid sequence of the Mr 10,000 delta zein. The NH_2-terminus of the mature protein begins at the upper left and proceeds left to right. On the scale in the lower part of the figure the tall bars indicate positions of methionine residues and the short bars positions of the cysteine residues.

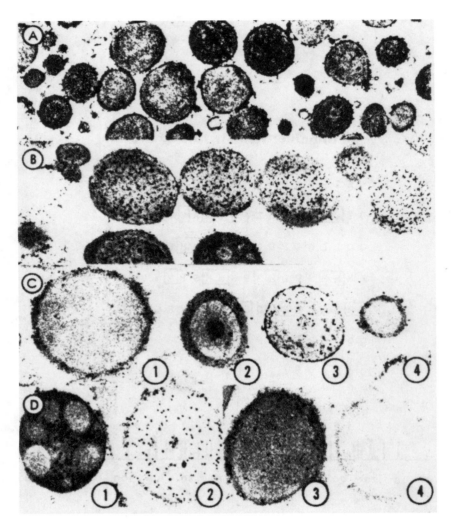

Figure 6. Localization of zeins in protein bodies by immunocytochemical labeling. Protein bodies were isolated from 18-day-old maize endosperm, prepared for electron microscopy and stained with immunogold complexes as previously described (Lending et al., 1988). Panel A shows a sample of protein bodies isolated by sucrose gradient centrifugation, stained with lead citrate and uranyl acetate. X 13,000, Bar = 1uM. Protein bodies in panel B were stained with immunogold using antibodies against the alpha zeins. Panel C, 1-4, shows selected protein bodies following staining with gamma zein antibodies, while panel D, 1-3, shows selected protein bodies following staining with gamma zein antibodies. X 55,000, Bar = 0.5 uM. Panel D, 4, illustrates staining with preimmune serum (from Larkins et al., 1988).

FOOD QUALITY • 75

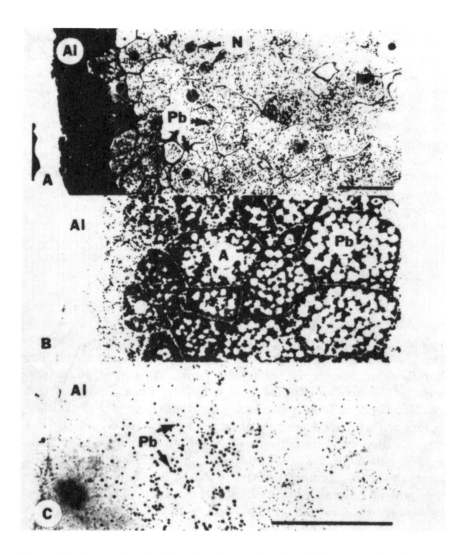

Figure 7. Immunolocalization of zein proteins in subaleurone cells of maize endosperm. Sections of 0.5 uM from 20 day-old W64A kernels were prepared and mounted on gelatin-coated glass slides. Sections were immunostained as previously described (Lending et al., 1988), except gold labeling was intensified using INTENSE II (Janssen Life Sciences Products, Olen, Belgium) according to the manufacturer's instructions. The section in panel A was stained with Toluidine Blue only (X 320). The section in panel B was immunostained with alpha zein antibody (X 320), while the section in panel C was stained with beta zein antibody (X 700). Abbreviations: Al, aleurone; PB, protein body; N, nucleus; A, amyloplast. Bar = 50 uM.

We used antibodies against the alpha, beta, and gamma zeins to localize these proteins with immunogold staining (Lending et al., 1988). Antibodies against the alpha zeins react primarily with the light-staining regions of protein bodies (Fig. 6B). Many protein bodies are uniformly labeled with this antiserum, while in others the colloidal gold is limited to light-staining pockets that are embedded in a dark-staining matrix. Immunogold staining with either the beta or gamma antiserum is complementary to that of alpha; the dark-staining regions are labeled with the colloidal gold particles (Fig. 6C, 1-4; 6D, 1-3). The dark-staining areas occur at the surface of the protein body or sometimes as central inclusions. Some protein bodies show essentially no labeling with the beta and gamma antisera, while some small protein bodies stain intensely. Reaction with preimmune serum reveals no immunostaining of protein bodies (Fig. 6C, 4).

The variation of zein composition in protein bodies results primarily from differences in the types of proteins synthesized in different regions of the endosperm. As illustrated in Fig. 7A, differences exist in cell size within the endosperm. Cells of the aleurone and subaleurone layers are generally smaller than those found in deeper regions of the endosperm. Staining of protein bodies with Toluidine Blue shows that the protein bodies in the subaleurone cells are smaller than those in cells deeper in the endosperm. Staining of similar sections with antibodies directed against the alpha zeins indicates that small amounts of these proteins are synthesized in the subaleurone region, while much larger amounts are synthesized in cells found deeper within the endosperm tissue (Fig. 7B). Staining of these same regions with antibodies against the beta zein (Fig. 7C) or gamma zeins (data not shown) is complementary to that of the alpha zein. In these cases, the protein bodies in the subaleurone layers appear larger and stain more intensely than cells in deeper regions of the endosperm. In cells that are a transition between the two regions, the outer part of the protein bodies stains more intensely with the beta and gamma antisera giving a donut-like appearance. At progressively deeper regions of the endosperm, staining with the beta and gamma zeins becomes more diffuse. This diffuse staining is consistent with the superficial distribution of the beta and gamma zeins on the surface of the protein body as revealed by electron micrographs showing immunocytochemical staining (Fig. 6).

These studies indicate that there are specific interactions that take place between the various types of zein proteins and, furthermore, that differences exist in the amounts of zeins synthesized in different regions of the endosperm. Thus, changing the structure of zein proteins may have an impact on the substructure of protein bodies and differentially affect various regions of the endosperm.

Genetic Engineering of Lysine-containing Alpha Zeins

All of the zein proteins are devoid of lysine, which is a nutritionally essential amino acid for monogastric animals. Because zeins account for more than 50% of the total protein, maize is of poor nutritional quality. Mutations have been identified that reduce the synthesis of zeins and thereby increase the proportion of lysine in the endosperm (Nelson, 1969). However, these mutants have proven to be agronomically inferior and consequently have not been utilized.

As an alternative method to improve the lysine content of maize seeds, we have investigated the potential for modifying the genes encoding zein proteins by site-directed mutagenesis. Using this approach, we hope to maintain a high protein content while at the same time increase the lysine content of the seed.

Initially, we modified a gene encoding an Mr 19,000 alpha zein so that lysine substitutions were introduced at various positions (Wallace et al., 1988). The primary amino acid sequence of this protein is illustrated in Fig. 8. The numbered and lettered positions in the sequence correspond to the mutant clones listed on the sides of the figure where single or double lysine substitutions (constructs 1-5), or short oligopeptides (constructs a-c), were inserted. At the position marked by the asterisk, a 17-kd peptide from the SV40 VP2 protein was inserted to make a more dramatic alteration in the protein structure. Our initial premise was that lysine substitutions in the regions outside the repeated peptides might be more easily tolerated than those within the repeats.

The best method to test whether these alterations perturb the synthesis and processing of the modified zein proteins into protein bodies would be to transfer the genes into a corn plant. But since the technology for such an experiment is not yet available, we used a heterologous system. In previous studies we showed that Xenopus laevis oocytes are able to translate zein mRNAs and process the proteins into protein bodies (Hurkman et al., 1981). We therefore transcribed the mutant clones to make synthetic mRNAs and injected these into Xenopus oocytes. To assay for protein body formation, we fractionated oocyte homogenates with metrizamide gradients and identified radioactively labeled zein proteins that sedimented in membrane vesicles at the density of maize endosperm protein bodies (Wallace et al., 1988).

Figure 8. Primary amino acid sequence in an Mr 19,000 alpha zein protein showing positions of lysine substitutions, oligopeptide insertions, and a large peptide insertion. The plasmid constructs listed at the sides of the figure indicate the site of the substitution or insertion. Synthesis of the mutant sequence and *in vitro* transcription of mRNA were previously described (Wallace et al., 1988).

When compared with protein bodies from oocytes injected with total zein mRNA (Fig. 9A), those injected with mRNA encoding the unmodified Mr 19,000 alpha zein were somewhat less dense (Fig. 9B). In both cases, the proteins sedimented in membrane vesicles within the range of densities observed for protein bodies from maize endosperm. Labeled zein protein that was added exogenously to the homogenate sedimented near the top of the gradient (Fig. 9C), indicating that the occurrence of zeins in dense regions of the gradient was not due to nonspecific interactions.

Modifications that introduced single lysine residues (Fig. 9D), two lysine residues (Fig. 9E-F), or a short oligopeptide insertion (Fig. 9G) yielded proteins that aggregated to form structures with densities similar to those formed by the unmodified zein protein (cf. Fig. 9B and Fig. 9D-G). However, the addition of the 17-kd fragment of the VP2 protein of SV40 in the NH2-terminal end of the zein prevented its aggregation; in this case the protein sedimented near the top of the gradient (Fig. 9H).

The fact that protein bodies can form within membranes of oocytes demonstrates that the association of zein polypeptides into a protein body is primarily a function of interactions between the proteins themselves once they enter the lumen of the RER. It appears that aggregation can occur with an alpha-type protein alone, although the protein bodies that form are slightly less dense than those formed from the mixture of zein polypeptides. We are unsure of the explanation for this and experiments to investigate the interaction of mixtures of these polypeptides are in progress. Experiments reported here indicate that the addition of lysine residues, or even lysine-rich oligopeptides, to the alpha zeins does not perturb their ability to aggregate. Nevertheless, more significant alterations such as the insertion of a large peptide can deleteriously affect their interactions.

Although the addition of lysine residues to zein polypeptides does not appear to significantly affect the synthesis and processing of these proteins, it remains to be demonstrated that these changes will be tolerated in a plant seed.

As a preliminary way to investigate this, we transferred zein genes into petunia and tobacco plants via a Ti-plasmid vector. Unfortunately, the promoters regulating zein gene expression do not appear to be properly regulated in transgenic petunia or tobacco plants (Ueng et al., 1988; Schernthaner et al., 1988). However, with a dicot promoter the zein coding sequences are transcribed and the mRNAs translated to give detectable quantities of protein (Williamson et al., 1988). We have therefore used this approach to produce the modified zein proteins in seeds of transgenic tobacco plants. Seeds produced by these plants are currently being analyzed to characterize the synthesis, transport, and stability of the modified zein proteins.

This research was supported by grants from the National Institutes of Health (GM 36970) and Lubrizol Genetics to BAL.

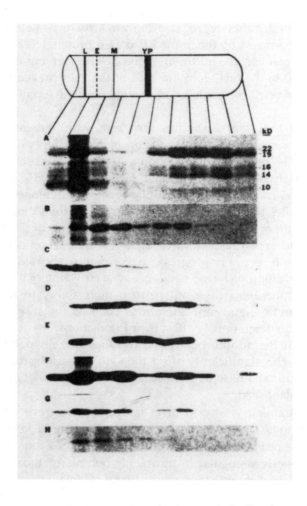

Figure 9. Density gradient separation of zein protein bodies from mRNA-injected oocytes. The oocytes were injected with synthetic mRNA transcripts and ^{3}H-leucine as previously described (Wallace et al., 1988). Following homogenization, extracts were analyzed by centrifugation in 4-ml gradients of 10-50% metizamide. Gradients were fractionated manually and the distribution of radioactive alcohol-soluble proteins determined by SDS-PAGE and fluorography. Panels correspond to the following mRNAs (see Fig. 7). A) native total zein mRNA; B) wild-type 19 kD alpha-zein; C) exogenous ^{3}H-zein added to oocyte homogenate; D) MZIK 32; E) MZ(NK 110-NK 159); F) MZi 32-33; G) MZi 127-128; H) MZ44-SV40. The Mr of the zeins is shown in panel A; other zeins migrated at Mr 19,000 except the MZ44-SV40 protein (H), which had an Mr of 35,000 (from Wallace et al., 1988).

References

Argos, P., Pedersen, K., Marks, M.D. and Larkins, B.A. 1982. A structural model for maize zein proteins. J. Biol. Chem. 257: 9984.

Esen, A. 1986. Separation of alcohol-soluble proteins (zeins) from maize into three fractions by differential solubility. Plant Physiol. 80: 623.

Hurkman, W.J., Smith, L.D., Richter, J. and Larkins, B.A. 1981. Subcellular compartmentation of maize storage proteins in *Xenopus* oocytes. J. Cell Biol. 89: 292.

Kirihara, J.A., Hunsperger, J.P., Mahoney, W.C. and Messing, J.W. 1988. Differential expression of a methionine-rich storage protein gene in maize. Mol. Gen. Genet. 211: 477.

Larkins, B.A. and Hurkman, W.J. 1978. Synthesis and deposition of zein proteins in maize endosperm. Plant Physiol. 62: 256.

Larkins, B.A., Pedersen, K., Marks, M.D. and Wilson, D.R. 1984. The zein proteins of maize endosperm. TIBS 9, 306.

Larkins, B.A., Lending, C.R., Wallace, J.C., Galili, G., Kawata, E.E., Geetha, K.B., Kriz, A.L., Martin, D.N. and Bracker, C.E. 1988. Zein gene expression during maize endosperm development. In: The Molecular Basis of Plant Development (Goldberg, R.B., ed.), Alan R. Liss (in press).

Lending, C.R., Kriz, A.L., Larkins, B.A. and Bracker, C.E. 1988. Structure and immunocytochemical localization of zeins. Protoplasma 143: 51.

Nelson, O.E. 1969. Genetic modification of protein quality in plants. Adv. in Agronomy. 21: 171.

Schernthaner, J.P., Matzke, M.A. and Matzke, A.J.M. 1988. Endosperm-specific expression of a zein gene promoter in transgenic tobacco plants. EMBO J. 7, 1249.

Pedersen, K., Argos, P., Naravana, S.V.L. and Larkins, B.A. 1986. Sequence analysis and characterization of a maize gene encoding a high-sulfur zein protein of Mr 15,000. J. Biol. Chem. 261: 6279.

Prat, S., Cortades, J. and Palau, J. 1985. Nucleic acid (cDNA) and amino acid sequences of the maize endosperm protein glutelin-2. Nucleic Acids Res. 13: 1493.

Shotwell, M.A. and Larkins, B.A. 1989. The biochemistry and molecular biology of seed storage proteins. In: The Biochemistry of Plants: A Comprehensive Treatise (Marcus, A., ed.). Academic Press (in press).

Ueng, P., Galili, G., Sapanara, V., Goldsbrough, P.B., Dube, P., Beachy, R.N. and Larkins, B.A. 1988. Plant Physiology 86, 1281.

Wallace, J.C., Galili, G., Kawata, E.E., Cuellar, R.E., Shotwell, M.A. and Larkins, B.A. 1988. Aggregation of lysine-containing zeins in protein bodies in *Xenopus* oocytes. Science 240: 662.

Williamson, J.D., Galili, G., Larkins, B.A. and Gelvin, S.B. 1988. Plant Physiol. (in press).

Genetic Modification of Traits of Interest to Consumers and Processors

David A. Evans
DNA Plant Technology Corporation
2611 Branch Pike
Cinnaminson, NJ 08077, USA

The raw material for many branded food products is purchased as a commodity. Food companies provide added-value based on processing or marketing. The tools of plant biotechnology, which shorten the time for crop improvement and permit development of novel germplasm, offer the food industry the opportunity to modify their raw materials. Such modifications will permit development of plant varieties specifically selected for traits with added value for the processor or the consumer. Biotechnology-developed varieties offer food companies the opportunity to have proprietary raw materials for use with specific brand names. The cell genetics tools of clonal propagation, somaclonal variation, gametoclonal variation, and protoplast fusion permit new variety development in an intermediate time scale, making them attractive for food industry applications. When integrated with conventional breeding, these intermediate-term technologies will permit modification of raw materials to meet food industry specifications over the course of the next several years. The longer term technologies of plant genetic engineering will continue to impact manipulation of specific traits resulting in second-generation products specifically designed for consumers and the food industry.

Introduction

The food industry has recently become active in the application of plant biotechnology. The biotechnology tools of cell genetics and molecular genetics permit new variety development on an accelerated schedule, making them attractive for food industry applications. When integrated with conventional breeding, the technologies discussed in what follows will permit modification of raw materials to meet food industry specifications during the next several years. Biotechnology-developed varieties offer food companies the opportunity to have proprietary raw material for use with specific brand names.

Cellular Genetics

Clonal propagation

Clonal propagation provides potential for the large-scale production of genetic carbon copies of superior varieties for commercial use. A wide variety of plant species can be clonally propagated from plant tissue using carefully defined culture media (Hu and Wang, 1983). Examples of commercially propagated tissue culture crops include strawberry, asparagus, and oil palm (Morris, 1983).

Future clonal propagation advances currently being researched focus on propagation using bioreactor technology for industrial-scale clonal propagation (Styer, 1985). Potential applications of clonal propagation include development of cloned varieties, propagation of elite plants and hybrids difficult or time-consuming to propagate using conventional seed technology, virus-free nursery stock, and propagation of certain perennials, estate crops, fruit trees, and forest trees.

The generation of clones can be achieved at both the organ and cellular level. For micropropagation (or mericloning), the apical meristem, the mass of cells at the growing tip of shoots, is removed and induced to develop many additional shoots. These can be separated and rooted, each one a clonal copy (Murashige, 1978). Alternatively, cells taken from the donor plant can be induced to reproduce in culture and then plants regenerated. The advantage of this second approach is that many cells, and therefore many plants, can be grown in a shorter period of time. From these cells, there are two routes to plant regeneration: organogenesis, in which the cells first form shoots and are then separated and rooted; and somatic embryogenesis, in which each cell develops into an embryo, identical to the seed embryo but arising from body or somatic cells. Each somatic embryo has a shoot and a root; thus, in one step both growing regions necessary for plant development are formed. Somatic embryos can be grown in liquid medium in large numbers (Ammirato, 1983).

Clonal propagation is proving particularly useful in plants where the unique combination of traits would be lost during seed production or where standard methods of asexual propagation are inefficient or lacking. Clonal propagation is also valuable in the reproduction of certain high-value hybrids that are too expensive to produce commercially due to the absence of male sterility, or where seed production is inefficient or seeds show poor germinability. It is especially useful in crops that must be propagated asexually due to natural infertility, such as of the root and tuber crops.

The propagation of an elite line of asparagus with resistance to the fungal diseases Fusarium and Verticillium has produced double the usual yields in field trials. Clonally propagated strawberries have proven hardier and produce substantially better yields. Clonal propagation technologies have been applied to potato for the propagation of elite, virus-free plants and to sugarcane for the introduction of specific variants arising in tissue culture. By manipulating development, microtubers have been formed in potato cultures, providing numerous units that can be efficiently delivered for planting in the field. A similar system is being developed for tropical yams.

Clonal propagation may also allow for the rapid development of improved plant varieties and replace the time-consuming conventional breeding procedures currently used for certain perennial varieties, such as estate crops, fruit trees, and forest trees. Clonal propagation may also permit the scale-up of production of certain superior perennial breeding lines selected either in the field or from biotechnologically altered plants for the rapid introduction of cloned varieties. Certain cloned varieties may have the potential to become competitive with seed-propagated varieties. For example, cloned estate crops may be an important potential commercial application of clonal propagation. Plant breeding of perennial crops has lagged far behind conventional crop breeding because of the time requirement for maturity. Using tissue culture it is possible to produce uniform, high yielding, and disease-resistant perennial crops. The methodology for cloning perennials has been developed in coffee, cacao, and oil palm trees.

Somaclonal variation
In contrast to clonal propagation, which faithfully produces genetic carbon copies, evidence has accumulated suggesting that regeneration of plants from callus, leaf tissue explants, or plant protoplasts (wall-less cells) results in recovery of somaclonal variants (Evans and Sharp, 1986). This allows the researcher to modify genes in cultured cells of existing plant varieties for development of new, improved varieties. Regeneration of plants from callus culture of existing commercial vari-

eties has been associated with recovery of genetic variants. In tomato, potato, wheat, corn, and other crops somaclonal variants have been produced and selected for development of new breeding lines that have moved plant breeding a step forward with new agronomic and processing benefits.

The genetic variability recovered in regenerated plants probably reflects both pre-existing cellular genetic differences and tissue culture-induced variability. Skirvin and Janick (1976) systematically compared plants regenerated from callus of five cultivars of geranium. Plants obtained from geranium stem cuttings *in vivo* were uniform, whereas plants from *in vivo* root and petiole cuttings and plants regenerated from callus were all quite variable. Changes were observed when regenerated plants were compared to parent plants for plant and organ size, leaf and flower morphology, essential oil constituents, fasciation, pubescence, and anthocyanin pigmentation. Long-term cell cultures result in tissue culture-induced variability with regard to chromosome number that is expressed in both callus and in plants regenerated from callus. This results in regeneration of aneuploid plants that are commercially useless in sexually propagated species. On the other hand, aneuploidy may not interfere with productivity of asexually propagated crops such as sugarcane and potato.

Somaclonal variation depends on the occurrence and recovery in plants regenerated from cell culture of Mendelian and non-Mendelian genetic variation. These changes in the integrity of the genome are attributed to mutant induction, mitotic crossing-over, and organelle mutation. Two important selection steps of the procedure serve as sieves to recover a population of plants produced by self-fertilization of first-generation tissue culture-derived plants that are most suitable for a breeding program: (1) the culture medium provides a sieve for singling out cells from the foundation cell populations which possess genome competence for plantlet regeneration; (2) greenhouse selection identifies those plants regenerated from cell cultures of the donor plant with normal development that are capable of undergoing flower and fruit formation and which set seed. This permits elimination of plants with deleterious genetic backgrounds. The regenerated plants are selfed and the resulting seed is used for field trial evaluation and selection of breeding lines.

Evans and Sharp (1983) made the first detailed genetic evaluation of the plants regenerated from cell culture of a sexually propagated crop plant. In evaluating the plants regenerated from tissue cultures of tomato, they have provided evidence for recovery of single gene changes using somaclonal variation. This discovery demonstrates that the biotechnology tool of somaclonal variation can result in stable changes in the genetic traits of tissue culture-regenerated plants. Several of these single-gene changes of tomato have been well characterized and localized to specific regions on the chromosome map of tomato. Among the variants of tomato observed using this procedure were changes in fruit color, plant architecture, and characteristics to improve the ease of mechanical harvesting. This new tool provides scientists with the ability to develop new breeding lines for the generation of improved varieties for food and industrial products from existing varieties within a reasonable time frame. This compresses the time requirement for modification of existing varieties.

Gametoclonal variation
Gametoclonal variation is recovered by regenerating plants from cultured microspore or pollen cells or pollen still contained within the anther (Morrison and Evans, 1988b). This genetic variation is the result of both recombination occurring during meiosis and tissue culture induced mutation. The gametes are products of meiosis, governed by Mendel's laws of independent assortment and segregation and contain half the amount of genetic information found in normal somatic cells. The most exciting opportunity is to culture anthers of hybrid plants that contain genetic information that represents a mixture of two parents. The resulting haploid plants can be used to produce new true-breeding recombinant lines that express the best characters of each parent.

To recover gametoclones, anthers containing immature microspores are removed from plants grown in the greenhouse or field and are transferred to a culture medium suitable for recovery of haploid plants. These haploid plants are then treated in some fashion (usually treatment with the chemical colchicine) to double the chromosome number to produce true-breeding lines with two copies of all genes. These doubled haploid plants are extremely valuable as breeding parents since a breeder can be certain that no masked genetic traits are present and can, therefore, predict the genetic behavior of these plants in subsequent generations (Sharp et al., 1984).

Gametoclonal variation has exciting commercial opportunities in two cases (Evans et al., 1984). First, many crops, particularly some cereals, can be regenerated more easily from anthers than from somatic tissue. Hence, anthers of rice and wheat have been cultured to recover new variants using a procedure resembling somaclonal variation. This approach has resulted in new breeding lines of rice with disease resistance.

Second, anther culture can be used to achieve hybrid sorting. That is, anthers of hybrid lines are cultured to recover haploid plants that express the best characteristics of each parent line. This method of hybrid sorting has been used extensively and quite successfully by Chinese workers to produce several new crop varieties. For example, two rice varieties, Hua Yu No. 1 and No. 2, were developed that express high yield, resistance to bacterial blight and wide adaptability. These and other rice varieties developed via gametoclonal variation are currently grown on more than 100,000 hectares in China. Similarly, new varieties of wheat (Jingdan 2288) and tobacco (Danyu No. 1) that simultaneously express multiple, desirable characters have been developed using gametoclonal variation (Chu, 1982).

The most exciting potential for gametoclonal variation has not yet been fully realized. Based on the success of the Chinese workers, it appears feasible to use gametoclonal variation to "sort out" commercial F1 hybrid seed. In one step it will be possible to recover true-breeding lines that express the most desirable characteristics of new commercial hybrids.

Protoplast fusion

Protoplast or plant cell fusion is a technology that permits the combination of different plant cells and leads to the production of new hybrid plants. In some cases hybrids can be produced that cannot be obtained using conventional breeding methods, resulting in new breeding lines.

Plant cells are surrounded by a cell wall that protects the cell's contents. This wall can be enzymatically digested to produce protoplasts, wall-less cells, that are easier to manipulate. Mixed protoplasts isolated from plant cells of two parents can be fused by using a multistep chemical treatment that includes treatment with polyethylene glycol. Fused protoplasts are then grown in culture medium suitable for plant regeneration to permit recovery of new hybrid plants.

This procedure is limited to those species that are capable of plant regeneration from protoplasts. Nonetheless, several somatic hybrid plants have been recovered that express genetic information from each parent line. This procedure has been particularly useful for recovering hybrids between tobacco and closely related species, but recently has been extended to several food crops (Evans and Bravo, 1988).

In most cases, somatic hybrids produced by protoplast fusion contain a fairly uniform mixture of genetic information from each parent line. Hence, it is necessary to integrate these hybrids into a conventional breeding program. The process is time-consuming but opens the door to the transfer of new traits that are not accessible using conventional breeding approaches.

While no new plant varieties have been produced using protoplast fusion, new breeding lines have been produced between cultivated tobacco and several wild tobacco species with disease resistance (Evans et al., 1981), and new male sterile lines of rapeseed have been developed in several laboratories.

Another use of protoplast fusion is the transfer of cytoplasmic genetic information. During fertilization, the male cytoplasm is preferentially excluded from the developing embryo. Thus, in most crops, conventional sexual hybrids contain and express only maternally derived cytoplasmic genes. As protoplast fusion combines both parent cytoplasms into a single protoplast, fusion can be used to produce unique nuclear-cytoplasmic combinations not possible using conventional breeding (Gleba and Evans, 1984). In some cases, even though protoplasts fuse, the two parental nuclei do not fuse. In these cases, cybrids (cytoplasmic hybrids) are produced that contain novel mixtures of chloroplast and mitochondrial genes. By creating cybrids, small blocks of DNA can be transferred between breeding lines using a one-step process. Several agriculturally important traits, such as disease resistance, herbicide resistance, and male sterility for hybrid seed production, are all encoded in chloroplasts and mitochondria. Hence, use of protoplast fusion to transfer organelles will open the door to development of many new plant varieties.

This technology has greatest application to those crops capable of plant regeneration from protoplasts. This list currently includes tobacco, tomato, potato, rice, corn, carrot, rapeseed, lettuce, and alfalfa, and is expanding rapidly.

Molecular Biology

Over the last several years, technology has been developed to stably introduce new useful genes into plants. Historically, two obstacles have hindered development of this technology: identification and isolation of useful genes; and transfer of such genes into higher plants. While some specific, useful genes have been isolated and cloned, there must be significant progress in this area before varieties with value-added consumer or processor attributes can be developed. On the other hand, tremendous progress has been made in transformation/regeneration in the past five years.

Must of the progress of gene transformation has been an outgrowth of studies on the bacterium *Agrobacterium tumefaciens*, responsible for crown gall disease in plants (Nester et al., 1984). The disease is caused by transfer of a segment of DNA from a bacterial plasmid into the nuclear genome of an infected plant. Manipulation and modification of this bacterial plasmid DNA, called T-DNA, has led to development of vectors that are capable of replication in *Escherichia coli*, transfer to *Agrobacterium*, and integration into plants (Horsch et al., 1988). It is possible to modify such vectors to introduce useful genes, selectable markers, and regulatory sequences.

A leaf disc transformation system based on *Agrobacterium* infection has been developed to easily transfer DNA between *Agrobacterium* and plant cells (Horsch et al., 1985). Several researchers have worked on novel methods to transfer any DNA (including disarmed *Agrobacterium* plasmids) into plant cells. The most promising methods include electroporation of protoplasts (Potrykus et al., 1985) and use of high-energy microprojectiles to transfer DNA (Wu et al., 1987). All of these methods require plant tissue culture regeneration to obtain transformed plants. One method, pollen transformation, has shown promise as a means to eliminate the need for tissue culture (Arntzen et al., 1988).

In most instances genes introduced using T-DNA are inherited as single dominant genes, thereby becoming a stable part of the plant genome. While limited data has been collected on field performance of transformed plants, it is expected that such plants represent a precise method to develop new varieties.

Protection of New Developments

The above techniques are all aimed at the development of new genetic variants of crop plants. But if plant biotechnology is to attract corporate interest, it is also necessary to protect new germplasm developed by these tools. The two approaches that are available to protect new plant breeding lines are male sterility for production of hybrid seed and molecular fingerprinting.

Male sterility is under genetic control and permits inexpensive and efficient production of hybrid seed. Protoplast fusion opens the door to efficient transfer of genes controlling male sterility, especially since these genes are encoded in mitochondria. Hence, cellular geneticists can work with the plant breeder to first identify superior parent lines to permit inexpensive production of the hybrid seed. Commercial F1 hybrids produced using male sterility are protected as growers that save seed of the F1 hybrid no longer have the uniform hybrid characteristics in their seed; F2 seed segregates and produces a non-uniform planting containing many undesirable plants. Hence, when growers use F1 seed, they must return year after year to purchase new F1 seed.

Molecular fingerprinting is the use of molecular methods to precisely characterize the genetic information of new breeding lines. DNA can be isolated from a breeding line and can be treated with restriction enzymes to generate a precise banding pattern that uniquely reflects the DNA present in a new breeding line produced by genetic engineering. Such precise characterization will aid in identification of varieties covered by the Plant Variety Protection Act or by Utility Patents to Plants.

Application of Technology to Food Products

Agronomic benefits
Improvement in agronomic benefits represent a short-term opportunity of biotechnology. In many instances, traits can be identified that are controlled by single genes and these genes can be easily isolated (Table 1). Agronomic improvement can result in lower cost of goods sold to food processors and can help to insure consistent raw material supply.

Table 1. Agronomic traits controlled by single genes

Herbicide resistance
    Sulfonylureas
    Imidazolonones
    Atrazine

Pest resistance
    Viral diseases
    Bacterial diseases
    Fungal diseases
    Nematodes
    Insect resistance (e.g., BT-gene)

One consumer attribute that can be addressed is pesticide residues. There is increasing consumer pressure to reduce the spraying of pesticides and the level of pesticide residues. Zero-pesticide residues is actually becoming a marketing issue in California, and is certain to spread to other parts of the U.S. and to the other developed countries in the near future. Virus resistant tobacco and tomato plants have been developed through genetic engineering. *Bacillus thuringiensis* produces a toxin that is inhibitory to insect attack. Scientists have been successful in transferring the BT toxin gene to plant colonizing bacteria and plants to achieve insect resistance.

Herbicide resistance plants of tobacco and tomato have been developed and are now being field tested. Within the next five years, we can probably expect large-scale testing of such varieties of rice, corn, soybeans, tobacco, and tomatoes.

Plants tolerant to drought conditions have been engineered. Scientists have selected isolates of *Pseudomonas* which do not synthesize the ice nucleation protein. These isolates are able to competitively colonize plant material, thus providing a degree of frost tolerance. These are a few examples of the types of agricultural products to enter the market in the near term.

Processing benefits
The modification of processing and functional attributes will be feasible as biotechnology offers the tools for establishing the required cor-

relations between genes and performance and the technology to precisely alter the genome of raw materials. Functional properties are those chemical and physical properties that affect the processing or the quality of the final food product. Food processors are increasingly recognizing the potential to modify these key attributes. The development of wheat for bread making is an example of plant breeders and processors working together closely to develop varieties with improved functional properties.

In many instances, food processors use essentially the same manufacturing processes—for example, all manufacturers of precooked rices prepare their rice for eventual rehydration by the consumer in the same manner. For processors to maintain the competitive advantage they have enjoyed in the past and to create new products, they will need to leverage their starting or raw materials. The use of parboiled rice as the starting raw material for this process illustrates the importance of the starch characteristics to the quality of the end product. Considerable variability in the chemical and physical properties of rice exists among the various genotypes of rice. Generally, the starch properties are closely tied to the geometric classification, and little attempt has been made to cross these boundaries. Farms of Texas (Alvin, Texas) in cooperation with DNA Plant Technology Corporation, DNAP, is developing varieties of rice which combine individual traits desired by the processors of rice, beer, soups, quick-cooking rice, and other products. Other examples of how genetically manipulatable traits in cereal grains that can affect the processing of foods include the effect of gluten content of wheat bread making, of starch properties on the texture of breakfast cereals, and of starch properties on the staling of baked goods.

Most of the commercial strawberry varieties have been developed for the fresh market with little attention, other than in the Pacific Northwest, given to such processing considerations as color, flavor, and freeze-thaw stability. To combine the fresh market and processing attributes into commercial lines would take considerable time and, depending upon the objectives, may not be attainable using conventional breeding. However, by supplementing conventional breeding with a thorough understanding of the biochemical processes and biotechnological approaches, it may well be possible to stack the desired freeze-thaw stability, excellent color, and flavor with the postharvest attributes required for the fresh market. There is evidence among existing germplasm that the freeze-thaw stability of strawberries can be dramatically affected.

Several programs are directed towards increased soluble solids in processing tomatoes. Over the years plant breeders have made steady progress towards this goal, which can greatly reduce the costs of processing the tomatoes into its many products. One can imagine the difficulty in enhancing a trait, which is multigene controlled and greatly affected by the environment. However, a combination of breeding and somaclonal variation made progress towards increased solids while still maintaining other agronomic and quality attributes. One PVPA certificate has been issued for a high-solids tomato variety developed by somaclonal variation (Evans, 1986).

Shelf life
The consumers' demand for year-round availability of fruits and vegetables and their willingness to pay a premium for quality will provide the incentive for biotechnologists to attempt to control ripening and senescence of fruit. Harvest quality of fruits and vegetables and the retention of quality at harvest will be greatly affected by biotechnology. Uniformity of ripening, uniformity of fruit size, as well as color and texture of fruit, will all be targets.

Polygalacturonase (PG) is the major enzyme responsible for the depolymerization of cell walls and the subsequent softening of the tissue during the ripening of fruits such as tomato (Huber, 1983). With the anticipation that reduced PG activity would result in improved shelf life and processing quality, scientists at the University of Nottingham and ICI have reported the inhibition of the expression of endogenous developmentally regulated gene for PG in transgenic tomato expressing antisense RNA (Smith et al., 1988). However, there was no observed reduction in softening as measured by compressibility. In the case of avocado, a fruit that does not ripen while attached to the tree, cellulase probably plays a major role in the solubilization of the cell wall polymers (Biale and Young, 1971; Tingwa and Young, 1975). Further understanding of the interaction of ethylene and cellulase accumulation may offer other opportunities for the control of cell wall deterioration in specific fruit. The control of ripening through the use of tissue-specific promoters for the control of ethylene-mediated ripening and senescence processes could yield substantial rewards in the uniformity of ripening and postharvest stability. This would necessitate the control of ACC synthase or the ethylene-forming enzyme (EFE).

Scientists have biochemically characterized many of the sweet corn mutants (Shannon and Garwood, 1984). Using similar antisense technology, it may be possible to further refine and control these pathways for the development of corn with specific functional properties. Similar mutants have been noted in rice (Omura and Sath, 1984). Similar control in other cereal grains could provide functionally and compositionally distinct cereal products with varying levels of sugar and complex carbohydrates.

It may be possible to minimize the respiration rates in harvested fruits and vegetables through biotechnology. By combining a postharvest reduction in respiration and secondary metabolism with optimum packaging and transportation, the postharvest quality and stability of produce can be maintained at a higher level. Browning of whole and prepared fruits can be controlled by controlling the enzymatic activity or availability of the substrate chlorogenic acid. There is considerable variation among genotypes of apples and peaches with regard to browning susceptibility (Mayer and Harel, 1981). This has been attributed to the differing levels of polyphenol oxidase, as well as substrate levels. The sulfur notes in *Brassica* vegetables become objectionable after a period of storage. These vegetables have limited shelf life and are subject to packaging conditions. It is anticipated that *Brassica* varieties with reduced respiration rates in the florets and reduced capacity to produce sulfur notes are achievable. There is already considerable variation in the shelf life and sulfur aromatics of broccoli.

Quality attributes
As with breeding, quality attributes of raw materials can be dramatically modified with biotechnology or the complementation of biotechnology with traditional genetics. Color, specific flavor components, carbohydrate fractions responsible for texture and water binding properties, and sugar levels can all be affected by this new science.

Variation has been reported in quality attributes of several crops, such as the weight of garlic (*Allium sativum*) cloves (Novak et al., 1982); the oil content of rapeseed (*Brassica juncea*) (George and Rao, 1983); the fruit color of tomato (*L. esculentum*) (Evans and Sharp, 1983); the sucrose content of sugarcane (*Saccharum sp.*) (Sreenivasan and Jalaja, 1983). DNAP was recently granted a PVPA certificate for a seedless bell-type pepper (Morrison and Evans, 1988a).

Color and flavor deficiencies greatly reduce the quality of the majority of tomatoes. Through the enhancement of pigment accumulation and of key flavor components and more strict control of the processes of ripening and senescence, biotechnology will improve the quality of tomatoes. Therefore, we can expect the development of new varieties of plants used for food and beverage with modified compositional characteristics to achieve better appearing and better tasting raw and processed products. There is also the opportunity to reduce levels of specific components through very targeted genetic manipulation. The major coffee processing companies are evaluating this new technology as a means of altering the caffeine concentrations of coffee.

Nutrition
There is tremendous opportunity to develop varieties with improved nutritional and physiological and even medicinal properties. This opportunity is exhibited in the considerable variability in vegetable, fruit, and cereal crops with regard to nutrient properties and functional carbohydrate fractions. The medical research profession is now attributing anticarcinogenic properties to ß-carotene, reduced cholesterol to increased levels of soluble fiber, and reduced incidence of heart disease to Omega-3 fatty acids. Traditionally, benefits such as antiviral, antithrombotic, diuretic, etc., have been attributed to specific secondary metabolites in a range of spices.

In 1985 a series of articles discussed the effects of Omega-3 fatty acids (linolenic, eicosapentaenoic, docosahexaenoic) in decreasing the incidence of coronary artery disease and lowering serum triglyceride levels in patients with hypertriglyceridemia. Studies indicated that eicosapentaenoic acid and docosahexaenoic acid were associated with reduced incidence of coronary heart disease. Essential fatty acids consisting of derivatives of linoleic and linolenic acids are responsible for the biosynthesis of prostoglandins, which are associated with blood vessels, joints, and skin (Kretchner and Kretchner, 1987). The oil of canola (*B. napus*) exhibits relatively high concentrations of linoleic and linolenic acid and contributes no cholesterol to the diet. Canola oil also contributes a very low level of saturated fats. This has spurred activity by several biotechnology companies to enter the developing multibillion-dollar oil industry by developing varieties of canola with improved levels of oil and modified fatty acid content.

The consumption of cruciferous vegetables has been associated with a reduction in the incidence of colon and stomach cancers (Birt, 1987). The American Cancer Society and the National Research Council recommend increased consumption of the cruciferous vegetables as a means of reducing the risk of some forms of cancer. Investigations into the active components has focused on flavonoids, indoles, and isothiocyanates. As the active compounds are confirmed, it can be expected that there will be an emphasis to develop varieties with increased levels of such components and to develop varieties exhibiting better flavor, texture, and shelf stability to take advantage of the increased demand.

Evidence from epidemiological studies have suggested that higher intake of carotene may reduce the risk of cancer (Colditz, 1987). Should raw materials with increased levels of carotenes be developed? Carrots (*Daucus carota*), for example, are thought to be the most important source of provitamin A carotenes and at the same time are a source of high-quality fiber (Robertson et al., 1979). The development of carrots with improved flavor and shelf stability, as DNAP's VegiSnaxR, fits current consumer trends towards the "good-for-you foods." Carrot selections with up to 12% sugar have been noted (Stommell and Simon, 1988). This would enhance the appeal of carrots and increase their consumption.

Recently, the high concentration of soluble ß-glucans of oats (*Avena sativa*) has been implicated in a reduction of cholesterol. There has been the suggestion that corn exhibits similar attributes. The technologies of plant breeding and biotechnology could lead to the development of varieties with higher concentrations of soluble fiber.

Health benefits have been attributed to specific secondary metabolites in a range of herbs (Duke, 1985). *Allium spp.*, for example, have a long history of use and folklore pertaining to their ability to inhibit specific diseases. Other herbs have been associated with a diversity of remedies, generally attributed to secondary metabolites.

Obviously, there will be the development of new varieties with improved health benefits, but what could eventually evolve from these activities directed toward the improvement of health benefits is a new field of medical foods. Key metabolites which exhibit physiological and medicinal attributes will be enhanced in new varieties or perhaps produced via cell culture as an ingredient or as an over-the-counter drug. The American Heart Association has indicated that it will endorse certain types of foods, a position which they have avoided in the past. This may provide the opportunity and the incentive for industry to develop varieties with these enhanced properties. In the future, biotechnology will allow the combination of traits previously unattainable.

Conclusion

Plant biotechnology directed towards the development of improved plant varieties for the food processor or for the consumer represents a tremendous opportunity. It is now possible to begin to breed the raw plant material according to end product or processor specifications. In the short term, this will lead to varieties with agronomic improvements resulting in cost-savings benefits. In the long term we can expect significant improvements in processing attributes, shelf life, quality attributes, and nutrition of our food products.

References

Ammirato, P.V. 1983. The regulation of somatic embryo development in plant cell cultures: Suspension culture techniques and hormone requirements. Bio/Technology 1: 68.

Arntzen, C.J., Evans, D.A., O'Dell, J.T., deBonte, L.R. and Loh, W.-H.T. 1988. Pollen mediated gene transformation in plants EPO. Appl. No. A2-0- 275-069.

Biale, J.B. and Young, R.E. 1971. The Avocado Pear. In: "The Biochemistry of Fruits and Their Products," p. 1. Part I, Academic Press, New York.

Birt, D.F. 1987. Anticarcinogenic factors in cruciferous vegetables. In: "Horticulture and Human Health," p. 160. Prentice Hall, Englewood Cliffs, New Jersey.

Chu, C.C. 1982. Haploids in plant improvement. In: "Plant Improvement and Somatic Cell Genetics," p. 129. Academic Press, NY.

Colditz, G. A. 1987. ß-carotene and cancer. In: "Horticulture and Human Health," p. 150. Prentice Hall, Englewood Cliffs, NJ.
Duke, J.A. 1985. Handbook of Medicinal Herbs. CRC Press, Inc., Boca Raton, Florida.
Evans, D.A. 1986. DNAP-9 Tomato. PVPA Certificate 84 00146.
Evans, D.A. and Bravo, J.E. 1988. Agricultural Applications of Protoplast Fusion. In: "Plant Biotechnology," p. 51. Univ. of Texas Press, Austin, TX.
Evans, D.A., Flick, C.E. and Jensen, F.A. 1981. Incorporation of disease resistance into sexually incompatible somatic hybrids of the genus *Nicotiana*. Science 219: 907.
Evans, D.A. and Sharp, W.R. 1983. Single gene mutations in tomato plants regenerated from tissue culture. Science 221: 949.
Evans, D.A. and Sharp, W.R. 1986. Applications of somaclonal variation. Bio/Technology 4: 528.
Evans, D.A., Sharp, W.R. and Medina-Filho, H.P. 1984. Somaclonal and gametoclonal variation. Amer. J. Bot. 71: 759.
George, L. and Rao, P.S. 1983. Yellow seeded variants in *in vitro* regenerants of mustard (*Brassica juncea*) Coss. var. RAI-5). Plant Sci. Lett. 30: 327.
Gleba, Yu.Yu. and Evans, E.A. 1984. Hybridization of somatic plant cells: Genetic analysis. In: "Genetic Engineering: Principles and Methods," p. 175, Vol. 6. Plenum Press, New York.
Horsch, R.B., Fry, J., Hoffmann, N.L., Wallroth, M., Eichholtz, D., Rogers, S.G. and Fraley, R.T. 1985. A simple and general methods for transferring genes into plants. Science 227: 1229.
Horsch, R.B., Fraley, R.T., Rogers, S.G., Klee, H.J., Fry, J. and Hinchee, M.A.W. 1988. *Agrobacterium*-mediated gene transfer to plants: problems and prospects. In: "Plant Biotechnology," p. 9. Univ. of Texas Press, Austin, Texas.
Hu, C.Y. and Wang, P.J. 1983. Meristem, shoot tip and bud culture. In: "Handbook of Plant Cell Culture," p. 177, Vol. 1, MacMillan Press, New York.
Huber, D.J. 1983. The role of cell wall hydrolases in fruit softening. Hort. Rev. 5: 169-219.
Klein, T.M., Wolf, E.D., Wu, R. and Sanford, J.C. 1987. High-velocity microprojectiles for delivering nucleic acids into living cells. Nature 327: 70.

Kretchner, P.J. and Kretchner, N. 1987. The contribution of horticultural crops and the role of Omega-3 fatty acids in health. In: "Horticulture and Human Health," p. 108. Prentice Hall, Englewood Cliffs, New Jersey.

Mayer, A.M. and Harel, E. 1981. Polyphenol oxidases in fruits—changes during ripening. In: "Recent Advances in the biochemistry of fruits and vegetables," p. 161. Academic Press, New York.

Morris, C. 1983. Developments in genetic engineering. Food Engineering 55: 18.

Morrison, R.A. and Evans, D.A. 1988a. Bell Sweet Pepper. PVPA Certificate 87 00124.

Morrison, R.A. and Evans, D.A. 1988b. Haploid plants from tissue culture: New plant varieties in a shortened time frame. Bio/Technology 6: 684.

Murashige, T. 1978. The impact of plant cell culture on agriculture. In: "Frontiers of Plant Tissue Culture," p. 15. Univ. Calgary Press, Calgary, Alberta, Canada.

Nester, E., Gordone, M., Amasino, R. and Yanofsky, M. 1984. Ann. Rev. Plant Physiol. 35: 387.

Novak, F.J., Harel, L. and Dolezel, J. 1982. *In vitro* breeding system of Allium. In: "Plant Tissue Culture," p. 767. Int. Assoc. Plant Tissue Culture, Maruzen, Tokyo.

Omura, T. and Sath, H. 1984. Mutation of grain properties. In: "Biology of Rice," p. 293. Japan Sci. Soc. Press, Amsterdam.

Potrykus, I., Shillito, R.D., Saul, M.W. and Paszkowski, J. 1985. Direct gene transfer: State of the art and future potential. Plant Mol. Biol. Reporter 3: 117.

Robertson, J.A., Eastwood, M.A. and Yeoman, M.M. 1979. An investigation into the dietary fiber content of named varieties of carrot at different development stages. J. Sci. Food Agric. 30: 388.

Shannon, J.C. and Garwood, D.L. 1984. Genetics and physiology of starch development. In: "Starch," p. 25. Academic Press, New York.

Sharp, W.R., Reed, S.M. and Evans, D.A. 1984. Production and application of haploid plants. In: "Contemporary Bases for Crop Breeding," p. 347. Pergamon Press, London.

Smith, C.J.S. et al. 1988. Antisense RNA inhibition of polygalacturonase gene expression in transgenic tomatoes. Nature 334: 724.

Sreenivasan, J.V. and Jalaja, N.C. 1983. Sugarcane varietal improvement through tissue culture. In: "Plant Cell Culture in Crop Improvement," p. 371. Plenum Press, New York.

Stommell, J.R. and Simon, P.W. 1988. Recurrent selection for total dissolved solids and sugar type in carrot (*Daucus carota* L.). HortScience 23: 121.

Styer, D.J. 1985. Bioreactor technology for plant propagation. In: "Tissue Culture in Forestry and Agriculture," p. 117. Plenum Press, New York.

Tingwa, P.O. and Young, R.E. 1975. The effect of Indole-3-acetic acid and other growth regulators on the ripening of avocado fruits. Plant Physiol. 55: 937-940.

Omega-3 Fatty Acid Improvements in Plants

Scott Bingham
David Kyle
Richard Radmer
Martek Corporation
Columbia, MD, USA

During the past several years there has been an increasing interest in the use of Eicosapentaenoic Acid (EPA, 20:5) and Docosahexaenoic Acid (DHA, 22:6), the putative health-enhancing components in fish oil, as food additives by the food industry. Oilseed crops with the ability to produce these long-chain polyunsaturated fatty acids (PUFAs) could provide a means to address this potentially lucrative market.

Higher plants do not normally generate PUFAs larger than 18 carbons. In contrast, some marine microalgae naturally synthesize EPA and DHA, and can be induced to accumulate large quantities of one or both of these PUFAs under appropriate conditions. Evidence to date suggests that the introduction into an oil-seed plant of the algal enzymes leading to EPA synthesis could utilize existing plant fatty acid precursors and generate EPA in the oil seed. Thus a relatively direct route exists for the development of a bioengineered EPA-producing oilseed plant.

Introduction

Coronary heart disease and cancer are the two major killers of Americans today. Nevertheless, these diseases are almost nonexistent among three-fourths of the world's population. Coronary heart disease is considered a disease of the affluent and, according to the National Heart, Blood and Lung Institute (NHLBI), the three major contributing factors to the disease are elevated serum cholesterol, high blood pressure, and cigarette smoking. These factors, for the most part, can be controlled by dietary and behavioral modifications in lifestyle. Epidemiological studies of incidences of coronary heart disease in Greenland Eskimos and coastal Japanese have suggested that the rarity of the disease in those populations is correlated with the consumption of substantial quantities of fish, whale, and seal, which are rich in omega-3 PUFAs (Bang and Dyerberg, 1972; Kromhout et al., 1985). The diets of Americans, on the other hand, are rich in the cholesterol and the saturated fatty acids associated with animal fats or the omega-6 PUFAs of vegetable oils. Several clinical investigations have compared the effects of diets rich in marine oils to those rich in vegetable oils under controlled conditions (Phillipson et al., 1985; Harris et al., 1983) and, in most cases, normolipidemic subjects with a fish oil-supplemented diet exhibited a significant decrease (30-60%) in plasma cholesterol and triglycerides. Hyperlipidemic patients exhibited even greater reductions in plasma cholesterol and triglyceride levels (50-80%). Furthermore, the plasma lipoprotein cholesterol which declined with the fish oil-enriched diet was exclusively the VLDL- and LDL-cholesterol.

Although less is known concerning the epidemiology of fish-enriched diets and cancer, recent studies sponsored by the National Cancer Institute (NCI) have indicated that high-fat diets predispose women to breast cancer (Davis, 1985; Karmali, 1987). Furthermore, studies with tumorous rats have shown that animals whose diets were enriched in fish oils (omega-3 PUFAs) exhibited a cessation of tumor growth, whereas tumors continued a rapid growth escalation in animals whose diets were enriched in corn oil (omega-6 PUFA). Several laboratories have shown that EPA specifically kills human breast, lung, and prostate cancer cells *in vitro* at concentrations which do not adversely affect normal fibroblasts (Begin et al., 1985; Booyens et al., 1984). These specific cytotoxic effects may be attributed to either a fluidization or destabilization of the inherently unstable membranes of malignant cells (Booyens et al., 1985), or to an inability to synthesize prostaglandin E_1 and/or thromboxane A_2 (due to a loss of delta-6 desaturase activity, a common observation in malignant transformation). In the latter case, it was noted that malignant cells can be "normalized" by providing gamma linolenic acid or EPA (Horrobin, 1980).

The initial findings discussed above indicate that there is an important connection between the EPA content of the diet and reduced incidence of coronary heart disease and many types of cancer. Consequently, there is a growing importance in understanding the biochemical bases for these observed effects. If additional clinical trials and tests confirm these reports, then the omega-3 PUFA consumption should become important to the general consumer, and EPA could be considered a valuable health-care product or food additive.

The only available source material for dietary supplements or clinical/biochemical investigations at the present time is the oil of certain marine fish. Some fish which are rich in omega-3 PUFAs (i.e., Menhaden) are restricted from use as food sources in many areas because of high level of PCB contamination. Cod liver oil, which is rich in both EPA and DHA, is also rich in the fat soluable vitamins A and D, and potential overdoses of these vitamins might occur if cod liver oil is ingested in clinically significant amounts relative to the PUFAs. Furthermore, the two omega-3 PUFAs in fish oil, EPA and DHA, are in approximately the same ratio in all fish. Although preliminary indications suggest that EPA may be clinically more important than DHA (Lee et al., 1985), it has been difficult to separate the effects of these inherently unstable and structurally similar fatty acids. There is clearly a need for an alternative source of these fatty acids in triglyceride (oil) form, as well as in the pure fatty acid or ethyl ester form.

At present, there are no confirmed reports that higher plants are capable of generating omega-3 PUFA larger than eighteen carbons. However, many species of phytoplankton, as well as a few fungi, have been reported to produce large amounts of EPA (Cohen, 1986; Loeblich and Loeblich, 1978; Shimizu et al., 1988). These organisms could provide an effective and valuable source of genetic determinants specifying omega-3 production, which could potentially be transferred to a seed-bearing crop plant. Although fatty acid biosynthesis is complex and the addition of EPA to a plant's suite of fatty acids would require transfer of multiple genes, the production of EPA as a plant vegetable oil component appears to be a legitimate long-term goal.

PUFA Biosynthesis in Higher Plants

Fatty acid biosynthesis in higher plants is complex, and appears to involve the participation of several different synthetase/desaturase systems. A schematic diagram of the relevant biosynthetic steps is shown below (from Jaworski, 1987; see also Stymne and Stobart, 1987):

C2 —> 16:0-ACP —> 18:0-ACP —> 18:1 —> 18:1-CoA —> 18:3-MGDG

Palmitic (16:0), stearic (18:0), and oleic (18:1) acids are synthesized by a nonassociated fatty acid synthase system that is acyl carrier protein (ACP) dependent. This system is localized in the chloroplast in green tissue and in proplastids in oil seeds (Stumpf and Shimikata, 1983). An important product of the system is steroly-ACP, which is desaturated to oleoyl-ACP and subsequently hydrolyzed to free oleic acid. This free fatty acid can then be transported to the cytosol and incorporated into a lipid such as phosphotidylcholine (PC) or monogalactosyldiacylglycerol (MGDG). These lipids can then serve as the substrate for microsomal elongation and desaturation systems, whose products can vary depending upon tissue and species.

Plants synthesize fatty acids in seeds to use as energy reserved for germinating seedlings. Although a given genetic strain will generally have identical oil composition through succeeding generations, the feasibility of altering the oil composition in storage lipids has been repeatedly demonstrated. There are wide variations in the fatty acids of storage lipids found in nature, from 10:1 fatty acids of elm trees to 22:1 fatty acids of *Nasturtium*. A summary of the fatty acids present in the seed oils of some common crop plants is given in Table 1.

Table 1. Fatty acid composition of some selected plant oils (%)

	Saturated				Unsaturated			
	14:0	16:0	18:0	Others	16:1	18:1	18:2	18.3	20:1
Corn	—	12.	2.5		—	29.	56.	.5	—
Olive	—	12.	2.	20:0	2	70.	13.	.5	—
Palm	1.	45.	3.8		—	40.	10.	.2	—
Peanut	—	11.5	2.	20,22,24	—	52.	27.	—	1.5
Rapeseed	—	5.	2.		—	63.	20.	9.	1.
Soybean	.5	12.	4.	12:0	—	25.	52.	6.	—
Sunflower	—	8.	3	20:0,22:0	—	20.	67.8	.5	—

Common Names

14:0	Myristic
16:0	Palmitic
18:0	Stearic
16:1	Palmitoleic
18:1	Oleic
18:2	Linoleic
18:3	Linolenic
20:1	Eicosenoic

Adapted from Zabriskie et al. (1980)

Note that a relatively few fatty acids make up the bulk of the seed oils in these plants, and that there are no PUFAs larger than 18 carbons in any of these plant seed oils. This is in accord with the general observation that high plants do not normally synthesize PUFAs such as EPA and DHA.

The fatty acid profiles of several important oilseed plants have been altered through plant breeding to generate cultivars such as high-stearate soybean (Nielsen, 1987) and high-oleic sunflower (Andreasen, 1987), presumably without altering their agronomic suitability. The opportunity that we are addressing here is unique in that its objective is to create new oilseed crop plants containing substantial quantities of long-chain omega-3 PUFAs (such as EPA) that are not normally synthesized by the plant.

PUFA Biosynthesis in Diatoms

Many oleagenous marine microalgae contain large quantities of PUFAs, and the relative abundance of each fatty acid can be considered a fingerprint for a particular species. For example, most species of diatoms contain large amounts of EPA and little or no DHA, whereas dinoflagellates have predominantly DHA (Loeblich and Loeblich, 1978). Since many of these species also produce lipids as a storage product, they can be considered natural producers of omega-3 PUFAs.

Table 2 is a compilation of the fatty acid profiles of three representative diatoms. Note that, in contrast to oilseed plants, the 18-carbon fatty acids, both saturated and unsaturated, are present in very low quantities. Equally striking is the high proportion of EPA [20:5(5,8,11,14,17)] in these organisms. One could speculate that this long-chain PUFA is being synthesized at the expense of the 18-carbon fatty acid pool, which would help explain the low proportion of these oilseed products in the diatoms.

Table 3 is a compilation of the results of a screening experiment carried out in our laboratory in which the total amount of extractable fatty acid and EPA content of several different diatom species were assessed. Note that there was a large variability in these two parameters from species to species, and that even within a species different culture conditions (i.e., nitrogen sufficient vs. nitrogen deficient) resulted in significant changes in the fatty acid profiles. The values of % fatty acid in the biomass in several cases compare favorably with the typical values of 20-30% of the dry weight generally observed for most commercial oilseeds, particularly in light of the fact that these oilseed values translate into values of less than 1% on a total plant biomassbasis. Thus it is clear that diatoms can be significant producers of fatty acids and EPA, and that they can provide the model and possibly the genetic information for the creation of EPA-producing oilseed plants.

Table 2. Fatty acid composition of some selected diatoms (%)

Fatty Acid	Double-bond Positions	*Navicula pelliculosa*	*Cyclotella cryptica*	*Phaeodactylum tricornatum*
14:0		3	5	9
16:0		16	24	11
16:1	9 (primarily)	31	30	27
16:2	6,9 & 9,12	10	6	14
16:3	6,9,12	7	4	10
16:4	6,9,12,15	0	0	3
18:0		1	2	<1
18:1	9 (primarily)	2	3	5
18:2	9,12	1	2	<1
18:3	6,9,12 & 9,12,15	1	1	<1
18:4	6,9,12,15	0	0	<1
20:4	5,8,11,14, & 8,11,14,17	0	0	<1
20:5	5,8,11,14,17	26	21	18
22:1	?	0	0	<1
22:5	4,7,10,13,16 & 7,10,13,16,19			

Adapted from Wood (1974)

Little is known about PUFA biosynthesis in diatoms, particularly as it pertains to EPA. Mead and Willis (1988) constructed a "consensus pathway" on the basis of studies in several other systems.

$$18:3(9,12,15) \longrightarrow 18:4(6,9,12,15) \longrightarrow 20:4(8,11,14,17) \longrightarrow EPA$$

According to this scheme, linolenic acid (presumably as a component of PC or MGDG) is enzymatically desaturated by a delta-6 desaturase to the corresponding 18:4 lipid. The 18:4 fatty acid (possibly as the acyl-CoA) is then elongated to 20:4. Subsequent enzymatic oxidation via a delta-5 desaturase results in the formation of a lipid containing EPA.

Table 3. Fatty acid and EPA contents of selected microalgal species

Species	%fatty acid in biomass	% EPA in lipid	% EPA in biomass
Cyclotella cryptica	10.6	23.8	2.5
C. cryptica, N deficient	29.6	10.0	3.0
Phaeodactylum tricornatum	9.2	26.9	2.5
P. tricornatum, N deficient	21.3	11.4	2.4
Cylindrotheca fusiformis	24.4	7.2	1.8
Navicula pelliculosa	25.4	9.0	2.3
Nitzschia angularis	7.7	24.7	2.2

Adapted from Behrens et al. (1988)

According to Pohl (1982), the biosynthesis of arachidonic acid [20:4(5,8,11,14)] and (presumably) EPA can take place by (at least) two distinct pathways. Starting from the common intermediate linoleic acid [18:2(9,12)], the pathways are as follows:

18:2(9,12) —> 18:3(6,9,12) —> 20:3(8,11,14) —> 20:4(5,8,11,14) —> EPA

18:2(9,12) —> 20:2(11,14) —> 20:3(8,11,14) —> 20:4(5,8,11,14) —> EPA

The upper pathway, often called the "animal pathway" or "gamma-linolenic acid pathway," has been found in several different algal classes (*Rhodophyceae*, *Chrysophyceae* and *Cyanophyceae*). This pathway is characterized by the desaturation of linoleic acid by a delta-6 desaturase. The resulting linolenic acid is then elongated before being desaturated to arachidonic acid by a delta-5 desaturase.

The lower pathway, which has been reported in Euglena, involves the direct elongation of linoleic acid. The resulting product is then sequentially desaturated by delta-8 and delta-5 desaturases to form arachidonic acid.

Although it is clear from the above discussion that the exact pathway of EPA synthesis in diatoms is still quite speculative, the fact remains that these organisms can and do synthesize large quantities of this long-chain PUFA. This EPA, at least in some diatoms, is present in both the polar and non-polar lipids (Behrens et al., 1988); consequently, we can expect that this fatty acid is produced as part of the general lipid metabolism, and does not reflect an unusual physically isolated pathway.

Equally important from our perspective is the likelihood that the microalgal EPA is derived from precursors present in large quantities in major oilseed crops, such as rapeseed. Thus these organisms could provide the source of the genetic material required to bioengineer an EPA-producing oilseed plant.

Genetic Engineering of Fatty Acid Biosynthesis: A Modest Proposal

It is clear from the data of Table 2 and the accompanying discussion that members of the plant kingdom (i.e., microalgae) exist that have a significant proportion of their fatty acids as long-chain omega-3 PUFAs. Furthermore, the EPA content of these plants can be increased by strain selection and manipulation of culture conditions (Behrens et al., 1988). Unfortunately, these are not crop plants, or even higher plants. Nonetheless, the fact that these microplants can produce EPA by what appears to be a simple extension of an existing higher plant pathway suggests that an EPA-producing oilseed plant could be developed. The data, discussion and speculation presented in the preceding two sections suggest that intriguing possibility of genetically combining the EPA-producing capability of microalgae with the proven agronomic success of oilseed crops. This could result in an oilseed plant with the lipid-synthesis characteristics of an alga and the agronomic characteristics of the plant into which the alga-derived EPA biosynthetic pathways were introduced.

The preferred approach to develop such an oilseed plant would be to choose organisms for gene isolation whose gene products are likely to both utilize existing plant fatty acid precursors and whose activities complement existing enzyme activities of the recipient plant. For example, if the consensus pathway of Mead and Willis (described above) is shown to be the primary pathway in diatoms, the most straightforward approach would be to introduce three genes into the recipient oilseed plant, namely those coding for the delta-5 and delta-6 desaturases and the elongase enzyme that generates 20-carbon fatty acid derivatives from the corresponding 18-carbon precursors.

Preconditions for Implementation

To achieve the aforementioned goals, a good deal of basic and applied research would be required. For example, in order to isolate the genetic determinants involved in EPA biosynthesis, one would first need to identify and purify the enzymes of interest. As discussed above, in the case of diatoms—which could be prime candidates because of their demonstrated ability to generate large quantities of EPA—there is little direct information in the literature concerning EPA biosynthesis. Thus, one would need to begin by identifying the enzymatic activities in cell homogenates using radiolabeled fatty acid precursors. These enzymatic activities would then need to be purified sufficiently to provide material for the generation of antibodies to the desired components. The antibodies would then be employed to screen a cDNA expression library constructed using the diatom mRNA. The clones so identified would be copies of the genes coding for the EPA enzymes. Each clone could subsequently be employed to isolate, for example, the native delta-5 and delta-6 desaturase genes and the elongase gene from a diatom genomic library. Concurrently, research in the targeted oilseed plant would be required to develop the systems for genetic transformation and gene expression.

Each of these research projects is well within the state of the art. However, a substantial committment of time and money would be required if one were to have some hope of overall success.

Conclusion

A successful execution of the program outlined above would provide the means for producing omega-3 PUFAs in conventional oilseed crops, thereby providing a new and economically important source of these compounds in vegetable oil. Parallel developments in the processing technology of these oils would provide a source of EPA-containing oil that would be compatible with foods in our normal diet. This scenerio could have a great influence on the health of each American, since it doesn't require the purchase of pills, or the recognition and treatment of a clinical disorder. Rather, it would represent a preventative medicine supplied to, and available to, the general public analogous to vitamin-supplemented foods.

References

Andreasen, T.J. 1987. Biotechnology for *Brassica* and *Helianthus* improvement. Presented at the World Conference on Biotechnology for the Fats and Oils Industry, Sept. 27 to Oct. 2, at Hamburg, West Germany.

Bang, H.O. and Dyerberg, J. 1972. Plasma lipids and lipoproteins in Greenlandic west coast Eskimos, Acta Med. Scan. 192: 85.

Begin, M.E., Das, U.N., Ells, G. and Horrobin, D.F. 1985. Selective killing of human cancer cells by polyunsaturated fatty acids. Prostaglandins Leukotrienes Med. 19: 177.

Behrens, P.W., Hoeksema, S.D., Arnett, K.L., Cole, M.S., Heubner, T.A., Rutten, J.M. and Kyle, D.J. 1988. Eicosapentaenoic acid from algae. In: "Biotechnology of Microbial Products" (in press).

Booyens, J., Engelbrecht, P., leRoux, S., Louwrens, C.C., vanderMerwe, P. and Katzeff, I.E. 1984. Some effects of the essential fatty acids linoleic acid and alpha linolenic acid and their metabolites gamma linolenic acid, arachidonic acid, eicosapentaenoic acid, decosahexanoic acid, and of prostaglandins A_1 and E_1 on the proliferation of human osteogenic sarcoma cells in culture. Prostaglandins Leukotrienes Med. 15: 15.

Booyens, J., Maguire, L. and Katzeff, I.E. 1985. Dietary fats and cancer. Med. Hypothesis 17: 351.

Cohen, Z. 1986. Products from Microalgae. In: "Handbook of Microalgal Mass Culture," p. 421, CRC Press Inc., Boca Raton, Florida.

Davis, J.M. 1985. Diet and breast cancer. National Fisheries Institute, Vol. 1, Issue 2.

Harris, W.S., Conner, W.E. and Murphy, M.P. 1983. The comparative reduction of plasma lipids and lipoproteins by dietary polyunsaturated fats: salmon oil vs. vegetable oils. Metabolism 32: 179.

Horrobin, D.F. 1980. The reversibility of cancer: the relevance of cyclic AMP, calcium, essential fatty acids and prostaglandin E_1. Med. Hypothesis 6: 469.

Jaworski, J.G. 1987. Biosynthesis of monoenoic and polyenoic fatty acids. In: "The Biochemistry of Plants," Vol. 9, p. 159, Academic Press, Inc., Orlando, Florida.

Karmali, R.A. 1987. Omega-3 fatty acids and cancer. In: "Polyunsaturated Fatty Acids and Eicosanoids," p. 222, Amer. Oil Soc. Publ., Champaign, Illinois.

Kromhout, D., Bosschieter, E.B. and Coulander, C. 1985. Inverse relationship between fish consumption and 20-year mortality from coronary heart disease. N. Engl. J. Med. 312: 1205.

Lee, T.H., Hoover, R.L., Williams, J.D., Sperling, R.I., RAvalese, J., Spur, B.W., Robinson, D.R., Corey, E.J., Lewis, R.A. and Austen, K.F. 1985. Effect of dietary enrichment with eicosapentaenoic and docosahexaenoic acids on *in virtro* neutrophil and monocyte leukotriene generation and neutrophil function. N. Engl. J. Med. 312: 1217.

Loeblich, A.R. and Loeblich, L.A. 1978. Division and Bacillariophyta. In: "Handbook of Microbiology," 2nd Ed. Vol. 2, p. 425. CRC Press Inc., West Palm Beach, Florida.

Mead, J.F. and Willis, A.L. 1988. The essential fatty acids: their derivation and role In: "CRC Handbook of Eicosanoids: Prostaglandins and Related Lipids," p. 85, CRC Press, Boca Raton, Florida.

Nielsen, N.C. 1987. Biotechnology for soybean improvement. Presented at the World Conference on Biotechnology for the Fats and Oils Industry, Sept. 27 to Oct. 2, at Hamburg, West Germany.

Phillipson, B.E., Rothrock, D.W., Conner, W.E., Harris, W.S. and Illingworth, D.R. 1985. Reduction of plasma lipids, lipoproteins and apoproteins by dietary fish oils in patients with hypertriglyceridemia. N. Engl. J. Med. 312: 1210.

Pohl, P. 1982. Lipids and fatty acids in microalgae. In: "CRC Handbook of Biosolar Resources," Vol. 1, Part 1, p. 383. CRC Press, Boca Raton, Florida.

Shimizu, S., Kawashima, H., Shinmen, Y., Akimoto, K. and Yamada, H. 1988. Production of eicosapentaenoic acid by *Mortierella* fungi. JAOCS 61: 1455.

Stumpf, P.K. and Shimikata, T. 1983. Molecular structures and functions of fatty acid synthetase enzymes. In: "Biosynthesis and Function of Plant Lipids," p. 1, ASPP, Rockville, Maryland.

Stymne, S. and Stobart, A.K. 1987. Triacylglycerol biosynthesis. In: "The Biochemistry of Plants," Vol. 9, p. 175. Academic Press, Inc., Orlando, Florida.

Wood, B.J.B. 1974. Fatty acids and saponifiable lipids. In: "Algal Physiology and Biochemistry," p. 236. Univ. of California Press, Berkeley, California.

Zabriskie, D.W., Armiger, W.B., Phillips, D.H. and Albano, P.A. 1980. Traders' Guide to Fermentation Media Formulation Traders Protein, Memphis, Tennessee.

Enhancing Meat Quality with Somatotropin (ST): Potentials, Limitations, and Future Research Needs

Norman C. Steele, Roger G. Campbell, Thomas J. Caperna and Morse B. Solomon
United States Department of Agriculture
Agricultural Research Service
Beltsville, MD 20705, USA
and
Animal Research Institute
Werribee, Victoria 3030, Australia

Growth hormone, or somatotropin, is a 191 amino acid residue protein which is normally secreted by the anterior pituitary gland and is associated with long bone growth and nitrogen retention. Recombinant DNA technology has made available vast quantities of species-specific growth hormone which emperically can function as a modulator/amplifier of gene expression toward the components of somatic growth, thereby hindering lipid accretion. For several reasons the swine industry will become the likely clientele for this technology. A species, such as the pig, with an inherent tendency toward excessive fat deposition, will more fully express genetic potential for lean tissue growth with strategic administration of porcine growth hormone. Based on the response of swine which differ in sex and fed diets of varying energy and protein contents, one could conclude ST technology is ideally suited for North American production practices. This technology could become commonplace in swine production once suitable sustained release delivery systems are developed and environmental variables are optimized for accelerated growth performance.

Introduction

Growth hormone (GH), somatotropin, and ST are synonyms for a single-stranded, 191 amino acid residue, nonglycosylated, protein containing two intramolecular disulfide bridges which is secreted by the anterior pituitary gland in response to positive (GRF [growth hormone-releasing factor]) and negative (STH [somatostatin]) effector secretions from the hypothalamus. GH has long been associated with growth processes (Walker et al., 1950), but only with the recent advances in recombinant DNA technology have sufficient quantities become available for long-term animal experimentation and to attract industrial interest toward product development for livestock application. In the United States, the livestock industry represents a $61 billion market (Agricultural Statistics, 1987) and, specifically, the $9.8 billion swine industry has been targeted as likely clientele for this technology. Due to productive efficiency currently constrained by physiological factors, the inherent tendency to synthesize and deposit excessive quantities of body fat and a very negative public image with respect to pork as a healthful meat product, the swine industry should benefit greatly from this technology. Several excellent references are cited (Machlin, 1972; Chung et al., 1985; Etherton et al., 1987; Etherton et al., 1986; Boyd et al., 1986) for background on the use of GH in swine. The purpose of this report is to describe the recent experiments with porcine GH (pGH) conducted at the Beltsville Agricultural Research Center. As designed, these projects did not represent efficacy tests of the technology, but were intended to provide data which might have global interest in swine production practices. Less pragmatically, we wish to quantify the absolute ceiling or limit for protein deposition by swine and pGH represents only a tool in pursuit of this goal.

Overview of Growth Biology

Growth of any species represents the accumulation of body weight due to hypertrophy of component tissues. Those tissues most critical for survival of a species have the highest priority for nutrient use and as such are supported even in marginal nutritional environments at the expense of tissues less critical for species survival. Sir John Hammond expressed this priority for nutrient use in a treatise describing the concept of nutrient partitioning (Hammond, 1952). Fig. 1 paraphrases the Hammond nutrient partitioning concept for growth of individual tissues and physiological function. On the abscissa a qualitative weighting of genetic and environmental determinants of tissue hypertrophy are proposed. According to this model, the elevation of extracellular nutrients occurs with increased plane of nutrition and neural and bone tissues extract with high-priority nutrients for maintenance and growth. Once differentiated, the hypertrophy of muscle and adipose tissue will occur in response to environmental factors, including nutrition, only to the degree permitted by genetic composition. Livestock production practices have focused on the maximum yield of lean tissue with modest accretion of body fat and within the specialty of animal nutrition one often assumes that genetic potential toward this end point is fully realized.

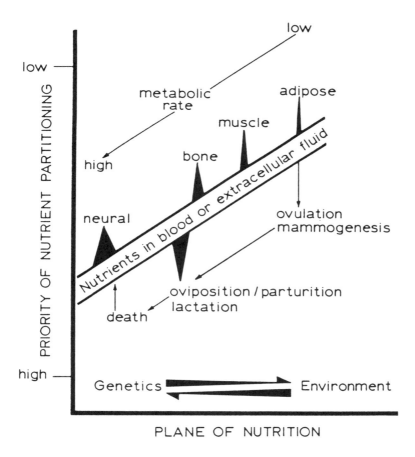

Figure 1. Hammond (1952) model of nutrient partitioning. Note that arrow boldness of a given tissue is indicative of the hierarchical need of the tissue for survival of the species and coincides with nutrient partitioning, or use, priority for maintenance and growth.

According to this model of nutrient partitioning, the relative accretion of muscle and adipose tissues could be altered either by increasing the slope of the line or altering the position of adipose tissue priority relative to muscle growth. These possibilities can be classified as partitioning (i.e., slope change) or repartitioning (i.e., relative position of component tissue growth) strategies. Partitioning strategies rely on genetic tools (gene optimization, gene amplification, and transgenic manipulations) to affect livestock performance. Repartitioning strategies rely on pharmacological, site-specific, alterations of metabolism. Examples of the latter include beta adrenergic agonists, thyroid-active compounds, and steroid implants.

The genetic tool most recognized for the modification of livestock growth is the gene optimization procedures used by animal breeders. A classical example is the Beltsville selection experiment for or against subcutaneous fat thickness (Davey et al., 1969) which simultaneously affected both muscle tissue deposition and skeletal growth. Recently a strain of swine has been characterized with an apparent infinite capacity for protein growth compared to typical commercial strains (Campbell and Taverner 1985, 1988). In this genotype, appetite is the only factor limiting protein accretion. These examples emphasize that gene optimization techniques can markedly influence both adipose tissue and muscle protein priorities for nutrient use. Such techniques as practiced over many centuries of livestock domestication are characteristically very slow and costly, but highly effective and generally regarded as safe with respect to environmental concerns. Unfortunately, the lack of knowledge regarding specific genes involved in growth and development processes and the inability to manipulate such genes within large populations of animals prevents rapid progress to reduce fat and increase lean tissue deposition using gene optimization procedures.

Recent progress in foreign gene insertion and expression (Hammer et al., 1985) has led to the production of agriculturally important "transgenic" animals. The promise of this procedure is that not only the slope of the nutrient partitioning priority line, but also the absolute genetic ceiling for both muscle protein and fat accretion can be altered by the only method capable of truly altering genetic composition. However, elucidation of mechanisms involved in the control of gene expression and the subsequent impact on growth and development as well as the remarkable negative public image of this technique will require considerable research investment and a lengthy education of the consuming public to be realized.

Yet another technique available to alter nutrient partitioning priority would be to preferentially increase the efficiency of muscle protein deposition, assuming that full genetic potential has been underestimated and constrained within normal physiological processes. In effect, this strategy involves "gene amplification" and describes those efforts which manipulate rate and composition of growth by ST administration. With strategic application of species-specific ST to induce supra-physiological concentrations of primary growth stimulants, those genes normally involved with growth processes are activated more completely and over longer periods of time, permitting a species to more fully express genetic potential for protein deposition. According to the Hammond model, pigs the size of steers, or steers the size of elephants, could not result from ST-induced gene amplification for protein deposition independent of proportionate increases in skeletal growth. However, pigs with feed efficiency comparable to broilers and dairy animals with 25% greater milk yields could enter the animal production system. This point is emphasized such that public concern regarding ST-induced growth alteration is not construed to be a violation of animal rights, specifically the genetic destiny of livestock.

An obvious example of naturally occurring gene amplification is the classic sex effect difference in rate and composition of gain comparing the intact male to the female pig (Campbell et al., 1985). Genetic composition does not differ greatly in this example, but expression of genetic potential for protein deposition is constrained in the female by the lack of androgen stimulation.

Porcine Somatotropin: Historical Information

Growth hormone has long been associated with long bone growth and nitrogen retention in laboratory animals (Li et al., 1948; Greenspan et al., 1949). Giles (1942) reported that injection of pigs with a pituitary-derived growth hormone preparation over a four-month period increased the width of the epiphyseal growth plate of long bone, but otherwise was without biological consequence. The biological potency of the GH preparation and lack of sufficient experimental detail prevents discussion for the lack of growth promotion. Turman and Andrews (1955) reported that a porcine pituitary preparation increased the efficiency, but not rate, of body weight accretion in swine. Significant increases of body water and protein content with a reduction in body fat was reported. Again, the biopotency of the GH preparation was unknown, but the authors noted that of those animals treated at a high dose (6 mg/15 kg body weight daily), mortality rate was 100%. In these pioneering studies much of the same potential as apparent in contemporary experiments for GH to alter nutrient partitioning in swine was noted, but hormone purity confounded the design of experiments and interpretation of results.

More recently, the research by Machlin (1972) and the efforts by Pennsylvania State University (Chung et al., 1985; Etherton et al., 1986; Etherton et al., 1987) and Cornell University (Boyd et al., 1986) investigators must be regarded as the basis for current interest in GH technology. Machlin found that treatment of growing pigs with a commercial preparation referenced to a rat tibia bioassay significantly improved rate and efficiency of body weight gain and the gain was generally of lean tissue mass and not lipid. In one experiment, GH treatment at levels of 220 and 1100 µg/kg body weight resulted in liver and kidney degeneration and a high mortality rate. Based on current information, doses used were far in excess for maximal biological benefit and reinforce that a dose-optimal does exist for pGH treatment.

With the resurgence of the national pituitary program in the early 1980's, Chung reported that daily administration of pGH (pituitary derived) at a dose level of 22 µg/kg body weight for a 30-day period improved rate of gain by 10%, efficiency of gain by 4%, increased muscle mass by 5%, and did not influence adipose tissue mass. Blood plasma glucose and insulin concentrations were increased by pGH while blood urea nitrogen concentration was decreased. This study was significant in that anabolic actions of GH were apparent in young pigs over a relatively short period of hormone treatment without induction of animal health complications. Subsequently, the optimal dose was tested by animal bioassay using levels of pituitary GH ranging from 10 to 70 µg/kg body weight daily and various production and carcass characteristics as end-point measurements. Rate and efficiency of gain improved over the doses tested and muscle mass continued to show fractional increase between a dose range of 35 and 70 µg/kg body weight.

Apparent in the Pennsylvania State University studies were the marked effects of pGH on nutrient partitioning in the pig and the potential for this technology to greatly alter swine production systems. However, of equal, if not greater, importance was the emergence of recombinant DNA technology permitting the mass production of species-specific $_r$pGH to provide a commercial product with marketing potential in the pork industry. Etherton and co-workers (1986) were the first to report that $_r$pGH, $_r$pST, was indistinguishable from pituitary-derived material in animal bioassay response. In a similarly designed dose-response titration experiment Boyd and co-workers at Cornell University reported that pituitary-derived pGH had optimal effect on most end-point parameters at a dose of 90 to 100 µg/kg body weight daily. Therefore, in two independent studies the dose optimal of pGH, natural or recombinant, is approximately 70 to 120 µg/kg body weight daily.

A consistent effect of pGH treatment is the reduction of voluntary appetite which in itself confounds the quantification of GH effects on carcass components from those effects attributable to feed restriction. In part, the greater improvement of feed efficiency compared to the effect on rate of gain would be anticipated considering the change in composition of gain (i.e., less fat and greater muscle protein and associated water). Several studies which have utilized either pituitary-derived (Wolfrom et al., 1986) or recombinant pGH (Kraft et al., 1986) found little if any benefit to growth and composition in pigs; however, nutrient intake, in particular protein content and composition, was marginal. Boyd and co-workers (personal communication) and recent efforts at Illinois (Easter, 1987) have revealed that protein intake can constrain realization of pGH benefit.

Porcine Somatotropin: Beltsville Studies

Within the discipline of nutrition one often loses the perspective that the mass and specific composition of the target tissue deposited defines the nutrient requirement of an animal. The diet with adjustments for obligatory losses associated with tissue deposition is merely the vehicle to supply the tissue requirement. With greater emphasis on lean tissue deposition and less lipid, the absolute and quantitative genetic potential for protein deposition is a very important concept in that this potential, or ceiling, defines the protein requirement of the pig. Underlying efforts at the Beltsville Agricultural Research Center was the thought of defining in quantitative terms the genetic potential for protein deposition. To this end, several studies have been conducted with designs to address this question. Simultaneously, several aspects pertinent to global swine production practices were addressed. Growth hormone within these experiments was used merely as a tool, albeit a very powerful tool, to maximize genetic potential for protein accretion.

Interaction of dietary energy intake and pST administration
Common in U.S. production practices is the use of ad libitum feeding systems which, to some extent, reflects the availability of grain resources. European feeding systems, in contrast, utilize limit-feeding due to the expense of grain resources, but additionally as a means to enhance lean tissue deposition. As reported (Etherton et al., 1987; Boyd et al., 1986), the beneficial effects due to pGH were confounded by ad libitum feeding and the reduction (10-15%) of voluntary appetite resulting from treatment.

Therefore, dietary energy was selected as the initial nutrient variable to reconcile possible synergistic effects as intake and pGH-altered nutrient partitioning. Additionally, Etherton and co-workers (1987) reported that benefit from pGH administration increased with increasing amounts of body fat. Shields et al. (1983) had previously reported that protein accretion rate was greatest in pigs between birth and approximately 55 kg live-weight and decreased slowly with further weight accretion. Therefore, the interaction of energy intake and pGH was examined over the live weight phase during which protein accretion is greatest (25 to 55 kg body weight).

Data in Table 1 summarizes the growth performance of barrows fed at one of three intake levels (AL- ad libitum, 80% AL, or 60% AL) a common diet and injected daily into the extensor neck muscles with excipient buffer or buffer containing pituitary-derived GH at a dose of 100 µf/kg body weight over the live-weight phase of 25 to 55 kg. Pigs were individually penned and feed intake and GH dose was adjusted weekly based on change of body weight. Consistent with previous reports (Etherton et al., 1987; Boyd et al., 1986), GH reduced voluntary appetite (10%) of pigs fed AL. Intake restriction masked this effect and animals consumed all feed offered regardless of GH dose. Growth rate increased linearly with increased feed intake and was amplified by 16 to 25% by GH treatment; however, feed efficiency was similar regardless of intake treatment and was consistently improved 23% by GH.

During the course of the experiment two animals were removed due to problems unrelated to treatment.

Table 1. Effects of energy intake and porcine growth hormone (pGH) administration on the growth performance of pigs from 25 to 55 kg live weight

Energy intake	pGH dose (µg/kg/d)	(No./trt.)	Feed intake (kg/d)	Energy intake (Mcal/d)	Daily gain (g)	Feed gain
Ad libitum (A)	0	(6)	2.32	8.10	905	2.57
	100	(6)	2.08	7.28	1052	1.96
80% Ad libitum (80% A)	0	(6)	1.64	5.72	670	2.45
	100	(6)	1.62	5.65	842	1.91
60% Ad libitum (60% A)	0	(5)	1.38	4.83	543	2.54
	100	(5)	1.34	4.70	681	1.95
SEM			.03	.09	22	.06
Model R^2			.98	.98	.93	.85
Significance of contrast, $P<$:						
0 pGH vs 100 pGH			.01	.01	.01	.01
A vs (80% A + 60% A)			.01	.01	.01	.31
80% A vs 60% A			.01	.01	.01	.27
pGH x [A vs (80% A + 60% A)]			.01	.01	.80	.65
pGH x [80% A vs 60% A]			.76	.73	.42	.67

Chemical analyses of carcass and visceral components permitted the calculation of accretion rates for water, protein, fat, and ash (Table 2) and using standard energy values, maintenance energy expenditure was estimated from the protein and fat components of growth. Body water, protein, fat, and ash all increased as a linear function with feeding level and, with pGH treatment, protein accretion increased by 35 to 49% while fat accretion was decreased by 25 to 32%. To emphasize this effect, pigs fed 40% less feed and treated with pGH (60% AL) exhibited protein accretion rates comparable to pigs fed AL and excipient-treated. Maintenance energy expenditure estimated at total energy retention of zero was increased by 17% due to pGH. This increased energy expenditure is due to the increase of lean body mass (i.e., metabolic rate of muscle > metabolic rate of adipose tissue) and a possible direct thermic effect of GH.

Table 2. Effects of energy intake and porcine growth hormone (pGH) administration on the rates of deposition of water, protein, fat, and ash and the fat:protein ratio in the empty body of pigs at 55 kg live weight

Energy intake	pGH dose (µg/kg/d)	Water	Accretion rates, g/d:			Fat:Protein
			Protein	Fat	Ash	
Ad libitum (A)	0	416	110	283	20.4	2.58
	100	567	151	193	36.8	1.28
80% Ad libitum (80% A)	0	310	85	171	15.4	2.04
	100	498	127	127	22.2	.99
60% Ad libitum (60% A)	0	279	78	110	11.9	1.41
	100	388	105	76	17.1	.73
SEM		18	4	11	2.2	.13
Model R^2		.88	.92	.89	.77	.85
Significance of contrast, P<:						
0 pGH vs 100 pGH		.01	.01	.01	.01	.01
A vs (80% A + 60% A)		.01	.01	.01	.01	.01
80% A vs 60% A		.01	.01	.01	.05	.01
pGH x [A vs (80% A + 60% A)]		.91	.26	.01	.01	.05
pGH x [80% A vs 60% A]		.05	.02	.64	.69	.15

Related to the mechanism of GH action, several studies have suggested that GH acts primarily on adipose tissue to decrease basal and insulin-stimulated lipogenesis (Walton and Etherton, 1986; Walton et al., 1987). Thereby, energy normally utilized to support fat accretion is available for protein synthetic processes. Data from this energy study revealed that the rate of lipogenesis as indicated by fat accretion rate is identical comparing excipient to pGH-treated pigs, but that the rate of body lipid turnover is increased. The increased feed energy required to maintain a unit of body fat results in a net reduction of body fat deposition. These data circumstantially imply that GH acts primarily on muscle tissue because if muscle protein effects were secondary to those on fat tissue the slope of the line describing energy retained as fat should be reduced compared to the slope for excipient-treated animals; this was not observed.

The following list describes conclusions from the dietary energy x pGH experiment:

- GH is a limiting physiological factor for lean tissue deposition in the young pig. Accretion of body fat reserve is not a prerequisite to demonstrate benefit from pGH administration.

- GH markedly increases the rate of protein deposition, regardless of energy intake and, therefore, pGH effects as reported are independent of and additive to the effects of such treatment on energy nutriture. The magnitude of benefit from pGH administration will be more fully realized with ad libitum feeding systems.

- As a qualification to the above statement, pGH increases maintenance energy expenditure by 17% and energy intake must be adjusted accordingly to realize full benefit from pGH treatment.

- Changes in rate and efficiency of growth attributed to pGH administration are direct consequences of stimulating protein deposition and energy associated with protein deposition combined with increased maintenance expenditure denies energy supportive of lipogenesis.

Interaction of dietary protein intake and pGH administration
In a recent study, the interrelationships between protein intake and pGH responsiveness was examined. Young crossbred barrows were used and dietary protein was varied between 11 and 27%. The diets were prepared to be isocaloric (3.8 Mcal DE/kg feed) and each formulation, regardless of protein content, contained the same lysine concentration (4.9 g lysine/Mcal DE). Thus, the influence of protein intake could be evaluated independent of lysine which is the primary limiting amino acid for growth of pigs. Soybean meal provided the variable source of dietary protein and was diluted with corn starch to yield five diets containing 27, 23, 19, 15 and 11% crude protein. The rations were limit-fed daily (approximately 80% restriction vs. ad libitum) to eliminate confounding effects of altered energy intake on interpretation of results. Recombinant pGH ($_r$pST) was injected daily (100 µg/kg)

into five groups of six pigs (i.e., one group on each of the five dietary regimes).

Table 3. Effects of protein intake and porcine growth hormone (pGH) administration on growth performance and selected carcass characteristics of pigs.

Dietary protein (%)	$_r$pGH μg/kg BW	Initial weight (kg)	Final weight (kg)	Daily gain (g)	Feed Gain	Loin eye area (cm^2)	Average backfat (cm)
11	0	32.3a	51.8	464	3.39	19.4	1.66
11	100	32.1	56.0	569	2.90	21.5	1.24
				(23%)[b]	(14%)	(11%)	(25%)
15	0	30.6	51.8	504	3.07	22.5	1.58
15	100	31.2	62.1	736	2.28	27.0	1.17
				(46%)	(26%)	(20%)	(26%)
19	0	30.9	53.4	536	2.93	21.9	1.66
19	100	31.3	62.9	760	2.23	28.0	1.09
				(42%)	(24%)	(28%)	(34%)
23	0	30.8	52.7	520	3.05	22.5	1.59
23	100	30.6	64.0	796	2.12	27.6	1.17
				(53%)	(30%)	(22%)	(20%)
27	0	29.0	49.8	496	3.02	20.4	1.21
27	100	29.4	60.7	744	2.15	27.8	0.99
				(50%)	(29%)	(37%)	(21%)
	SEM	1.3	2.1	33	0.12	1.2	0.15

[a] Values are means, n=6 pigs per treatment group.
[b] Values in parentheses represent percent change of pGH-treated pigs compared to excipient-treated controls.

An equal number of control pigs were treated with a similar volume of excipient buffer. Pigs were placed on treatment at 30 kg live weight and were treated for 6 weeks. Amounts of feed and pGH were adjusted weekly, based on change of body weight.

The purpose of this experiment was two-fold. First, to determine whether pGH could stimulate lean tissue growth in diets deficient (or marginally deficient) in protein and second, to determine whether the protein requirement is altered for pigs treated with GH. At present only preliminary growth performance data are available and are shown in Table 3. The data indicate that substantial changes in growth characteristics were evident at all levels of dietary protein intake as a result of pGH treatment.

Generally, it is apparent that at 15% protein (and higher) growth characteristics were quite similar. Furthermore, the improvement in performance of pigs treated with pGH and fed diets containing 15 to 27% protein diets were similar, at least in the parameters examined. In contrast, at 11% protein the magnitude of improvement in daily gain, feed:gain and loin eye area was found to be approximately 50% that observed at higher levels of protein intake. However, the net change in backfat due to pGH administration was similar at all levels of protein intake. This finding is in agreement with data from the previous study which demonstrated that aspects of fat metabolism were energy dependent and that the influence of pGH on protein and fat metabolism could be separated by nutritional manipulation. Furthermore, final interpretation of these data must await proximate analysis of the whole body and determination of protein and fat accretion rates in these animals. From the data presented here, it seems unlikely that the protein requirement for growing pigs is altered as a consequence of GH administration.

Effect of genotype on realization of pGH benefit
Virtually all reports of GH efficacy have utilized castrate males, barrows, which have the greatest rate of fat accretion and intrinsically the poorest rate of protein deposition. In addition, many countries market intact male pigs to take advantage of their superior growth characteristics. Therefore, genotypic effects were evaluated initially at the most fundamental level of genetic difference: animal sex. Subsequently, Campbell and Taverner (1988) repeated these efforts using pigs which differed not only in sex, but also realization of protein deposition potential through genetic optimization.

Data in Table 4 describe the response of female, castrate male, and intact male pigs fed a common diet and treated with either excipient buffer or pGH (100 µg/kg) for a 31-day period starting at 60 kg live weight. Apparent in the data was the superior rate (excipient buffer treated; male > female and castrate by 15%) and efficiency (male > female and castrate by 20%) of gain of intact male pigs. The sex effect difference was minimized with respect to rate of gain and eliminated with respect to feed efficiency by pGH administration. Chemical analysis of the components of growth confirmed that the vast improvement of feed efficiency observed in females and castrates was the result of accelerated protein and associated water deposition (Table 5). Therefore, GH technology will be of most benefit in those production environments which utilize pigs with the poorest protein accretion rates (i.e., female and castrate male pigs as typical in the U.S. swine industry). Furthermore, the lack of pGH effects of the same magnitude when administered to intact male pigs did not suggest that androgens and GH are additive stimulants of growth and development processes.

Table 4. Effects of animal sex and porcine growth hormone (pGH) administration on the growth performance of pigs from 60 kg live weight and treated for 31 days

Animal sex	pGH dose (μg/kgday)		Weight (kg)	Feed intake (kg/d)	Daily gain (g)	Feed:gain
		(No./trt.)				
Male (B; boar)	0	(6)	97.8	3.22	1186	2.72
	100	(6)	102.8	2.96	1342	2.21
Female (F; gilt)	0	(6)	91.8	3.38	1011	3.34
	100	(6)	98.8	2.73	1236	2.21
Castrate male	0	(6)	93.7	3.67	1057	3.46
(Ba; barrow)	100	(6)	98.1	2.84	1225	2.33
SEM			1.5	.13	43	.07
Model R^2			.53	.55	.57	.92
Significance of contrast, P<:						
0 pGH vs 100 pGH			.01	.01	.01	.01
M vs (F + Ba)			.01	.56	.01	.01
F vs Ba			.68	.13	.69	.09
pGH x [M vs (F + Ba)]			.81	.04	.59	.01
pGH x (F vs Ba)			.38	.50	.50	.96

The effect of pGH technology on strains of pigs which have diverged from commercial-type pigs with respect to growth characteristics was recently reported by Campbell and Taverner (1988; Table 6). Pigs of strain A represent a population with an infinite capacity for protein growth constrained only by appetite, while pigs of strain B are commercial-type animals typical of Australia. Intact male and female animals within strain were treated with either excipient buffer or pGH over the live weight phase of 60 to 90 kg and fed ad libitum a common diet. Consistent with the lack of GH x genotype interaction, pGH minimizes genetic differences. Within genotype, GH negates sex effects on the efficiency and composition of live weight gain.

Table 5. Effects of animal sex and porcine growth hormone (pGH) administration on the rates of deposition of water, protein, fat, and ash and the fat:protein ratio in the empty body of pigs treated for 31 days

Animal sex	pGH dose	Accretion rates, g/d:				
	(μg/kg/day)	Water	Protein	Fat	Ash	Fat:Protein
Male (B; boar)	0	467	196	317	23.6	1.40
	100	676	238	202	29.9	1.02
Female (F; gilt)	0	342	148	411	24.1	1.84
	100	680	235	185	37.8	1.04
Castrate male (Ba; barrow)	0	383	139	462	27.4	2.17
	100	615	225	223	29.2	1.22
SEM		29	9	27	1.9	.06
Model R^2		.84	.78	.79	.56	.91
Significance of contrast, P<:						
0 pGH vs 100 pGH		.01	.01	.01	.01	.01
B vs (F + Ba)		.01	.01	.01	.09	.01
F vs Ba		.54	.26	.07	.17	.01
pGH x [B vs (F + Ba)]		.12	.01	.01	.66	.01
pGH x (F vs Ba)		.07	.99	.73	.01	.23

With the original objective of defining protein accretion capacity in the pig, data in Table 5 has substantially refined the estimate. Previously, protein deposition capacity in the range of 180 grams/day has been considered an "ideal" reference (Whittemore, 1983). Regardless of animal sex, we report accretion rates for body protein between 235 to 250 grams/day. Considering the goal of animal agriculture to maximize protein output per animal unit, swine nutrition practices should adapt to satisfy this genetic potential. Considerable trade-offs exist in practical swine management, but the economic basis for these trade-offs must be recognized in the perspective that the genetic ceiling, or capacity, for protein growth in the pig is approximately 40% greater than previously estimated.

Table 6. Genotype and sex effects on the responsiveness of growing pigs to exogenous growth hormone administration (Campbell and Taverner, 1988)

	Strain:	A				B					
	GH dose:	0		100		0		100			
Item	Sex:	M	F	M	F	M	F	M	F	SEM	AOV*
daily gain, g		1177	894	1519	1250	992	737	1292	1080	17	G,S,GH
intake, kg/d		3.14	3.05	2.76	2.63	3.24	2.77	2.61	2.48	.32	S,GH
feed/gain		2.70	3.43	1.90	2.10	3.28	3.76	2.07	2.23	.24	G,S,GH,S-GH
backfat, mm		26.4	32.5	20.4	21.3	32.3	34.2	24.9	23.6	2.8	G,S,GH,S-GH
visceral wt, kg		7.8	7.4	7.5	8.5	7.4	7.1	8.4	8.6	.74	GH,S-GH
carcass lng., cm		78.2	76.2	80.0	77.9	77.1	77.3	78.3	79.6	2.2	GH,S-GH

*$P<.05$. Effects: G=strain; S=sex; GH= pGH; and, S-GH interaction.

Sustained effect of pGH administration

Ideally any manipulation of animal growth should involve those processes described as homeorhetic adaptations. Homeorhesis is defined as the chronic integration and coordination of metabolic processes to support a specialized physiological state (Bauman et al., 1982). Both for economy of GH used in the treatment of animals and ease of animal management, some advantage to the treatment of relatively small growing pigs could be realized if the pGH effects were sustained during the finishing phase of growth. In part, this challenges the homeorhesis theory of hormone action with respect to GH effects on growth. Assuming GH action is primarily exerted on those mechanisms controlling muscle protein deposition (i.e., satellite cell integration into myofibrils), decompensation of benefit following treatment withdrawal should not occur. This concept is also relevant should licensing approval mandate a lengthy withdrawal period following hormone treatment.

Barrows were treated daily with either excipient buffer or buffer containing pituitary-derived pGH (100 µg/kg body weight) over the live-weight phase of 30 to 60 kg. Pigs treated with excipient buffer were pair-fed to the intake noted in pGH-treated pigs to avoid energy consumption as a confounding factor. From 60 to 90 kg live weight no further injections were performed and all pigs were fed ad libitum. Growth performance segregated by live-weight period is summarized in Table 7.

Table 7. Effects of porcine growth hormone (pGH) administration from 30 to 60 kg live weight on the growth performance of pigs

pGH dose (μg/kg/d)	Live weight (kg)	Days	Feed intake (kg/d)	Daily gain (g)	Feed/gain
(Treated)					
0	30-60	39.8	1.98	798	2.84
100		29.5	1.94	1041	1.86
SEM		.6	.04	35	.04
Significance of contrast, P<:					
0 pGH vs 100 pGH		.01	.50	.01	.01
(Withdrawal)					
0 (0)	60-90	32.7	3.30	887	3.75
100 (0)		29.1	2.96	1020	2.92
SEM		.7	.09	31	.14
Significance of contrast, P<:					
0 pGH vs 100 pGH		.01	.02	.01	.01
(Overall)					
0	30-90	73.6	2.58	804	3.08
100		58.7	2.43	1009	2.41
SEM		.9	.04	12	.10
Significance of contrast, P<:					
0 pGH vs 100 pGH		.01	.02	.01	.01

Of the original 36% benefit in daily gain favoring pGH treatment, a 14% residual benefit was sustained during the finishing phase of growth. The 28% improvement of feed efficiency observed during the treatment phase deteriorated slightly (-6%) during the finishing phase of growth. Overall, treatment of young pigs with pGH accelerated growth velocity sufficient to reduce by 15 days the time interval to achieve 90 kg live weight utilizing 22% less feed.

Accretion rates of water, protein, and ash were significantly increased during the treatment phase, while carcass fat accretion rate was decreased, compared to excipient-treated control pigs (Table 8). During the withdrawal period, carcass water, protein, and ash deposition rates were significantly greater in pigs previously treated with pGH compared to excipient-treated, control animals. Fat deposition rate was comparable regardless of prior treatment.

Table 8. Effects of porcine growth hormone administration (pGH) from 30 to 60 kg live weight on the rates of deposition of water, protein, fat, and ash and the fat:protein ratio in the carcass of pigs

pGH dose (µg/kg/d)	Live weight (kg)	Water	Protein	Fat	Ash	Fat:Protein
(Treated)						
0	30-60	311	98	164	18.3	1.67
100		529	153	100	28.0	.65
SEM		53	16	27	3.7	.39
Significance of contrast, P<:						
0 pGH vs 100 pGH		.01	.01	.05	.05	.05
(Withdrawal)						
0 (0)	60-90	229	91	392	16.7	4.30
100 (0)		314	121	358	24.9	2.90
SEM		31	11	44	7.2	.72
Significance of contrast, P<:						
0 pGH vs 100 pGH		.01	.01	.50	.05	.01
(Overall)						
0	30-90	253	97	233	17.5	2.43
100		347	131	226	27.2	1.74
SEM		30	8	37	4.6	.43
Significance of contrast, P<:						
0 pGH vs 100 pGH		.01	.01	.75	.01	.01

Accretion rates, g/d (columns: Water, Protein, Fat, Ash, Fat:Protein)

These data indicate that, at minimum, the pGH benefit to somatic components of growth (i.e., water, protein, and ash) are maintained over control values during withdrawal and do not decompensate. The increased fat deposition of pigs previously treated with pGH suggested that the decrease in magnitude of stimulated protein deposition in effect created an energy surplus resulting in a large increase of lipogenic rate. Whether these data provide evidence for a homeorhetic adaptation is inconclusive, but at least the benefit derived from treatment of young animals with pGH results in a sustained benefit to growth velocity and lean body mass.

Conclusions

Somatotropin is a technology which could vastly alter the development pattern of livestock through amplification of genetic potential favoring the accretion of lean tissue and reducing the extent of fat deposition. Application of this technology in the swine industry should negate certain advantages in production practices of European countries, thereby making North American swine production more competitive in the world marketplace. Hindering application at this time is the development of a sustained-release drug delivery system; however, demonstration of a sustained effect in the young animal following a 30-day treatment regime may facilitate development. As this technology affects meat processing practices, competition among species with respect to market share for consumer selection, the impact on cereal grain utilization, and similar economic issues are beyond the scope of this report. Biologically, the effects of ST on reproductive performance, immune system function, progress in classical animal breeding practices, etc., are all researchable questions awaiting study.

Acknowledgement

$_r$pST was generously provided by IMC-Pittman-Moore, Terre Haute, IN.

References

Agricultural Statistics (Eds. L.D. Jewell and E. Johnson). 1987. United States Government Printing Office, Washington, DC 20402.

Bauman, D.E., Eisemann, J.H. and Currie, W.B. 1982. Hormonal effects on partitioning of nutrients for tissue growth: Role of growth hormone and prolactin. Fed. Proc. 41: 2548.

Boyd, R.D., Bauman, D.E., Beerman, D.H., DeNeergard, A.F., Souza, L. and Butler, W.R. 1986. Titration of the porcine growth hormone dose which maximizes growth performance and lean deposition in swine. J. Anim. Sci. 63 (Suppl. 1) :218.

Campbell, R.G., Taverner, M.R. and Curic, D.M. 1985. Effects of sex and energy intake between 48 and 90 kg live weight on protein deposition in growing pigs. Anim. Prod. 40: 497.

Campbell, R.G. and Taverner, M.R. 1985. Effects of strain and sex on protein and energy metabolism in growing pigs. In: P.W. Moe,

H.F. Tyrrell and P.J. Reynolds (Eds.), p. 78. Energy metabolism of farm animals. Europ. Assoc. Anim. Prod., Publ. 32 Rowman and Littlefield, USA.

Campbell, R.G. and Taverner, M.R. 1988. Genotype and sex effects on the responsiveness of growing pigs to exogenous porcine growth hormone (pGH) administration. J. Anim. Sci. 66 (Suppl 1): 257.

Campbell, R.G. and Taverner, M.R. 1988. Genotype and sex effects on the relationship between energy intake and protein deposition in growing pigs. J. Anim. Sci. 66: 676.

Chung, C.S., Etherton, T.D. and Wiggins, J.P. 1985. Stimulation of swine growth by porcine growth hormone. J. Anim. Sci. 60: 118.

Davey, R.J., Morgan, D.P., and Kincaid, C.M. 1969. Response of swine selected for high and low fatness to a difference in dietary energy intake. J. Anim. Sci. 28: 197.

Easter, R.A. 1987. Nutritional requirements and repartitioning agents. Proc. Univ. IL Pork Industry Conf., p. 193.

Etherton, T.D., Wiggins, J.P., Evock, C.M., Chung, C.S., Rebhun, J.F., Walton, P.E. and Steele, N.C. 1987. Stimulation of swine growth performance by porcine growth hormone: determination of the dose-response relationship. J. Anim. Sci. 64: 433.

Etherton, T.D., Evock, C.M., Chung, C.S., Walton, P.E., Sillence, M.N., Magri, K.A. and Ivy, R.E. 1986. Stimulation of pig growth performance by long-term treatment with pituitary porcine growth hormone (pGH) and a recombinant pGH. J. Anim. Sci. 63 (Suppl. 1): 219.

Giles, D.D. 1942. An experiment to determine the effect of growth hormone of the anterior lobe of the pituitary gland on swine. Amer. J. Vet. Res. 3: 77.

Greenspan, F.S., Li, C.H., Simpson, M.E. and Evans, H.M. 1949. Bioassay of hypophyseal growth hormone; the tibia test. Endocrinol. 45: 455.

Hammer, R.E., Pursel, V.G., Rexroad, C.E., Jr., Wall, R.J., Bolt, D.J., Ebert, C.M., Palmiter, R.D. and Brinster, R.L. 1985. Production of transgenic rabbits, sheep, and pigs by microinjection. Nature 315: 680.

Hammond, J.A. 1952. Physiological limits to intensive production in animals. Brit. Agric. Bull. 4: 222.

Kraft, L.A., Haines, D.R. and DeLay, R.L. 1986. The effects of daily injections of recombinant porcine growth hormone (rpGH) on growth, feed efficiency, carcass composition, and selected metabolic and hormonal parameters in finishing swine. J. Anim. Sci. 63 (Suppl 1): 218.

Li, C.H., Simpson, M.E. and Evans, H.M. 1948. The giantism produced in normal rats by injection of the pituitary growth hormone. III. Main chemical components of the body. Growth 12: 39.

Machlin, L.J. 1972. Effect of porcine growth hormone on growth and carcass composition of the pig. J. Anim. Sci. 35: 794.

Shields, R.G., Mahan, D.C. and Graham, P.L. 1983. Changes in swine body composition from birth to 145 kg. J. Anim. Sci. 57: 43.

Turman, E.J. and Andrews, F.N. 1955. Some effects of purified anterior pituitary growth hormone on swine. J. Anim. Sci. 14: 7.

Walker, D.G., Simpson, M.E., Ashling, C.W. and Evans, H.M. 1950. Growth and differentiation in the rat following hypophysectomy at 6 days of age. Anat. Record 106: 539.

Walton, P.E. and Etherton, T.D. 1986. Antagonism of insulin action in cultured pig adipose tissue by pituitary and recombinant porcine GH: potentiation by hydrocortisone. Endocrinology 118: 2577.

Walton, P.E., Etherton, T.D. and Chung, C.S. 1987. Exogenous pituitary and recombinant growth hormones induce insulin and insulin-like growth factor 1 resistance in pig adipose tissue. Dom. Anim. Endocrinol. 4(3): 183.

Whittemore, C.T. 1983. Development of recommended energy and protein allowances for growing pigs. Agric. Sys. 11: 159.

Wolfrom, G.W., Ivy, R.E. and Baldwin, C.D. 1986. Effect of native porcine growth hormone (npGH) injected intramuscularly in barrows. J. Anim. Sci. 63 (Suppl. 1): 219.

Summary

Stating a firm definition of "biotechnology" is a difficult task since the scope of this area changes dramatically with each new scientific development. Biotechnology integrates knowledge from several scientific disciplines to direct the metabolic potential of living systems in production of desirable materials.

Application of biological systems to the production and preservation of food and beverages has been with us since the early days of civilization. Initial discoveries and improvements in food production techniques were largely based on empirical observations. One of the early practitioners of biotechnology was Louis Pasteur. His integrated knowledge of chemistry and biology contributed significantly to enhanced food quality throughout the world (improved fermentation, control of spoilage organisms, and pasteurization). His ability to apply existing and new experimental knowledge to practical problems also laid the groundwork for the rapid development of microbiology as a distinct scientific discipline. This new field, coupled with ongoing advances in chemistry, physics, and biology led to the development of yet another field, molecular biology. These new fields have had major impact on biotechnology. Advances in all disciplines of biotechnology have increased the potential for increased diversity of food and medicinal goods as well as improved quality of raw commodities and final products. New products are already emerging and application of novel and traditional techniques will lead to significant advances in agricultural biotechnology as we enter the twenty-first century. The application of molecular genetics as a supplement to traditional plant and animal breeding techniques was highlighted in this session entitled "Genetic Engineering of Food Quality."

Dr. Brian Larkins discussed the distribution of major zein proteins in corn endosperm and their aggregation into protein bodies within the endoplasmic reticulum. One attempt to improve the quality of zein is to increase the content of the essential amino acid lysine, normally absent in zein. The experimental question was whether incorporation of a polar charged amino acid, such as lysine, would disrupt native zein functions, particularly aggregation into protein bodies. Through site-directed mutagenesis, single or multiple lysine codons were introduced into the gene encoding α-zein. Although the constructs could not be tested in corn (awaiting a suitable transformation/regeneration of α-zein to aggregate and form protein bodies in *Xenopus* oocytes. Furthermore, normal posttranslational processing occurred when modified zein genes were introduced into a dicotyledonous expression system. Thus a more nutritionally balanced endosperm should be possible when such mutations are introduced into corn. Dr. Larkins also highlighted efforts to develop high-lysine corn through a traditional maize breeding program. Although the high-lysine trait is stable and results in normal plump kernals, the exact protein which has been altered is unknown. Both studies suggest that development of nutritionally improved corn strains is both feasible and imminent.

Dr. Richard Radmer discussed the biosynthetic potential of certain brown marine algae to produce large quantities of long chain Ω-3 polyunsaturated fatty acids. Such fatty acids have received increased attention. Their dietary presence is correlated with reduced cholesterol levels and coronary disease in some populations. Although the hypotheses remains to be tested, ingestion of brown algae and accumulation of long chain polyunsaturates within the food chain may explain the high content of the eicosapentaenoic and docosahexaenoic acids in fish oils. Identification of the appropriate algal gene(s) and their incorporation and expression in oil-seed plants is an intriguing possibility which could result in large scale production of beneficial fatty acids. Dr. Radmer emphasized that the algal genetic and biosynthetic studies required for development of such transgenic plants are in the preliminary stages.

Dr. David Evans highlighted the variety of genetic tools which can be used in conjunction with traditional plant breeding techniques. Through manipulation of media and environmental conditions, cultured somatic or gametic plant cells can be induced to produce variations from the original stock. Genetic change can also be introduced via protoplast fusion between different cell lines. Protoplast fusion allows the nuclear genes of one cell line to be fused with the organelles of another. The organellar DNA often contains genes for key agricultural traits such as cytoplasmic male sterility, photosynthesis, pigment production, starch accumulation, and herbicide resistance. A third mechanism of introducing desirable genetic traits into plant cells is through recombinant DNA techniques. This last method is dependent on a thorough knowledge of the desired gene (or at least it's location in the parent genome) and current public concerns about environmental release of recombinant strains suggest that recombinant DNA techniques will be more fully utilized in the future.

The key to establishing whether new plant traits can be successfully established is regeneration of the cultured cells into mature reproductive plants. DNAP has been successful with plant regeneration in a large number of plant species including carrots, celery, corn, peppers, rice, tomatoes, wheat, several varieties of citrus, coffee, cucumber, melon, squash, rapeseed, and triticale. These regenerated plants can be crossed with existing breeding stock to yield plants with new properties (e.g. high solids tomatoes, uniform color carrots, large kernel popcorn, etc.). The potential exists to introduce a number of new genetic traits into traditional commodity vegetables. These traits can improve the efficiency of food processing, reduce the waste byproducts, and improve the nutritional value, flavor, texture, and pigment qualities of plant-based food products. Other molecular biological techniques, such as monoclonal antibodies can be harnessed for early detection of plant diseases. This approach may reduce the broad use of pesticides, and thus improve food quality.

Finally, Dr. Norman Steele shifted the discussion to the use of a somatotropin leading to improvement of meat quality in pigs. Traditional animal breeding practices have selected for swine that gain weight rapidly; price is determined on the basis of animal weight. However, a significant portion of the weight gain is due to fat accumulation. Current consumer concerns about cholesterol and saturated fat has had a negative impact on the pork industry; such levels of fat are increasingly unacceptable. Efforts to localize the structural and regulatory genes involved in growth and fat accumulation will require considerable study. Development of transgenic animals with improved "lean" characteristics will require additional effort as well as consumer acceptance. Dr. Steele summarized a number of experiments conducted at the Beltsville Agricultural Research Center. Porcine growth hormone (somatotropin) was used to enhance development of muscle and bone tissue, without stimulating a parallel increase in fat production. The result was a leaner animal (higher protein, water, and mineral content) exhibiting fairly rapid weight gain without rapid excessive fat accretion. Protein accretion rates between 235-250 grams/day could be attained compared to the previous "ceiling" of 180 grams/day. Differences in growth and protein accretion between sexes could be minimized through somatotropin treatment. The feed technique can be applied to improve the "lean" characteristics of pork without waiting for the eventual development and public acceptance of transgenic animals. Porcine somatotropin is now available in large quantities through recombinant DNA and animal cell culture techniques. The genetic potential for leaner meat is present in swine. Modulation of growth though growth hormone treatment can allow this potential to be achieved. However, the lack of efficient, economical delivery and application systems is a restraint which must be overcome for its use in lean pork production.

> Paul H. Tomasek
> *Department of Food Science*
> *Cook College*
> *New Brunswick, NJ, USA*

MOLECULAR COMPONENTS OF FOOD

Cell Wall Dynamics

Kenneth C. Gross
USDA/ARS
Product Quality and Development Institute
Horticultural Crops Quality Laboratory
Beltsville, Maryland 20705, USA

An understanding of the dynamic nature of plant cell wall structure and metabolism is requisite to controlling deleterious textural changes that occur after harvest of plant commodities. This paper describes our recent approaches and results on tomato fruit ripening along these lines. In assessing the regulation of ripening and the associated textural changes, we are looking at changes in cell wall composition, biosynthesis, and the potential involvement of cell wall fragments with elicitor activity in initiating ripening. We have partially purified a tomato wall fraction that, when added exogenously to mature green tomatoes, stimulates ethylene production (similar to results using galactose) and tissue breakdown. Also, radiolabeling studies have shown that wall biosynthesis continues throughout ripening. Thus, in addition to degradation of existing polymers, biosynthesis of modified types of cell wall polysaccharides may be involved in tissue softening. It is clear that a detailed structural and metabolic analysis of individual cell wall polymers, their potential biological activity, and how they are assembled in the wall, are requisite to controlling textural quality.

Introduction

The most economically important and well-studied area of cell wall biochemistry in relation to texture and food quality is the softening that accompanies fruit ripening. Fruit which have become overly soft are less desirable to consumers, and are more susceptible to mechanical damage and microbial decay during handling and storage. Tissue softening is thought to occur as the structural integrity of the wall is modified during ripening.

A number of excellent reviews have been written which deal with the intricacies of cell wall structure, methods of extraction, and structural analysis (Darvill et al., 1980; Dey and Brinson, 1984; McNeil et al., 1984; Selvendran and O'Neill, 1987; York et al., 1985). A detailed presentation of cell wall structure is outside the scope of this paper on plant cell wall dynamics in relation to food quality. It must be realized, however, that progress in regulating deleterious changes that occur to plant material after harvest hinges on a detailed analysis of, and development of ways of modifying or regulating the structure and metabolism of, the cell wall.

The plant cell wall has classically been separated into protein and three carbohydrate fractions: pectin, cellulose, and hemicellulose. The carbohydrate fractions comprise approximately 90-95% of the wall and protein about 5-10%. Pectic polysaccharides are mainly comprised of an α-1,4-galacturonan backbone with 2- and 2,4-linked rhamnosyl residues interspersed in the chain. It is thought that the 2,4-rhamnosyl residues may serve as branch points for neutral sugar side chains comprised primarily of arabinosyl and galactosyl residues. The galacturonan backbone can also contain methyl esters at the C-6 carboxyl or Ca^{++} bridges between adjacent carboxyls or between adjacent galacturonan chains. Hemicelluloses are polymers containing various neutral sugar residues such as xylosyl, glucosyl, fucosyl, and mannosyl, which may be covalently linked to pectin, and are noncovalently associated with cellulose via strong hydrogen bonding.

Most biochemical studies of textural changes have involved the cell wall related loss of firmness that accompanies ethylene-promoted fruit ripening. However, due to the multifarious nature of complex carbohydrates and to a lack of suitable methodology, research aimed at understanding the mechanisms which lead to textural changes have focused on degradation of cell wall pectins by polygalacturonase (PG; EC 3.2.1.15), an α-1,4-endo-D-galacturonase (DellaPenna et al., 1986; Hobson, 1964; Huber, 1983b). The concept is that textural changes occur as cell wall pectins are hydrolyzed by PG. This has been based on: 1) a general correlation, in some fruits such as tomato, between potential PG activity and softening; 2) an increase in soluble polyuronide during ripening; 3) the concomitant absence of appreciable levels of PG and softening in nonripening tomato mutants; and 4) the ability of crude PG preparations to hydrolyze isolated cell wall material *in vitro*; this evidence has recently been reviewed (Bartley and Knee, 1982; Brady, 1987; Huber, 1983b; Labavitch, 1981). However, it is readily apparent that other mechanisms must also be involved in regulating cell wall turnover and tissue softening. For example: 1) apples, strawberries, and tomato locule tissue soften in the absence of endo-PG activity

(Bartley, 1978; Huber, 1984); 2) some softening of tomato fruit occurs early in ripening prior to detectable PG activity (Brady et al., 1985; Hall, 1987), 3) the tomato mutant *dg* contains a normal level of PG but does not soften substantially (Jarret et al., 1984; Tong and Gross, 1988c); and 4) tomato fruit cell wall hydrolysis by PG may be limited *in vivo* by the number of PG binding sites or other factors (Brady et al., 1987). Thus, it is critical to study other mechanisms of wall polymer solubilization, as well as the structural characteristics of pectic polymer substrates that might regulate PG action *in situ*, such as calcium bridges, methyl esterification, and the presence of neutral sugar side chains. Consideration should also be given to other wall hydrolases and to the possibility that wall depolymerization depends on mechanisms other than hydrolytic cleavage. For example, there is evidence showing that pectin can undergo self-deaggregation; thus, non-enzymatic mechanisms of pectin depolymerization need to be considered (Fishman et al., 1986 and citations therein). Such possibilities may only become apparent as detailed studies of the structure and metabolism of cell wall polymers are conducted and realized.

Wall integrity may be varied, not only by the action of hydrolytic enzymes, but also by differential synthesis. Thus, a focus on the enzymes which are involved in cell wall polymer synthesis is appropriate. It will also be necessary to determine how the various wall polymers are affiliated in the wall. Because of the realization that some wall fragments released by hydrolytic agents have biological activity in surrounding tissues, its seems important to study this phenomenon in fruit ripening and softening as well. The objective of this paper is to describe our recent results on these matters using normal ripening and nonripening mutant (*rin*) tomatoes.

Experimental

Plant material
Tomato plants (*Lycopersicon esculentum* Mill.; cv. "Rutgers" and *rin*) were grown in a greenhouse under natural light. The *rin* (ripening inhibitor) mutant tomato line was isogenic to "Rutgers." Developmental and ripeness stages were based on days post-anthesis (DPA) and/or visually. Abbreviations are as follows: IM, immature green (18 DPA); MG, mature green (32-38 DPA); B, breaker (40 DPA); T, turning (41 DPA); P, pink (43 DPA); R, red (46 DPA). MG fruit were subdivided

based on internal appearance after harvest as follows: MG1, locules not jellied, some seeds cut; MG2, locules slightly jellied, few seeds cut; MG3, locules completely jellied, seeds not cut; MG4, locules jellied and yellow-orange in color.

Cell wall extraction and fractionation
Cell walls were extracted and fractionated as previously described (Gross, 1984). Briefly, outer pericarp from fresh "Rutgers" tomatoes was homogenized in 2 vol of 80% EtOH at -20°C using a Polytron, filtered through Miracloth, and sequentially extracted with 20 mM Hepes-NaOH (pH 6.9), porcine α-amylase (18 h at 37°C; Sigma Type I-A), phenol:acetic acid:ddH$_2$O (2:1:1, w/v/v), chloroform:methanol (1:1, v/v), and acetone. Filtration was with a sintered-glass filter. Fractionation involved sequential extraction of walls with CDTA (IAP; calcium chelator-soluble pectin), Na$_2$CO$_3$ (CBP; covalently bound pectin), KOH (HF; hemicellulosic fraction), and the residue was washed with ddH$_2$O (CF; cellulosic fraction). Wall fractions were stored over P$_2$O$_5$ at 37°C.

The CF was further fractionated by extraction with 8 M KOH containing 100 mM sodium borohydride under nitrogen at 25°C, yielding a second hemicellulosic fraction called HF II (Tong and Gross, 1988a). The extract was neutralized with glacial acetic acid (temperature kept below 35°C), and dialyzed for 72 h against ddH$_2$O at 4°C (Spectra-Por 1, 6-8 kD molecular weight cutoff). The remaining residue was washed with ddH$_2$O and lyophilized (CFII).

In vivo cell wall radiolabeling
Using a microsyringe, ^{14}C-sucrose was directly injected into the pedicel 1 to 2 cm above the calyx. All tomato fruit were injected with 5 µl ddH$_2$O containing 7 µCi of [^{14}C(U)]-sucrose (671 mCi/mmol) at 2:00 pm and harvested the following day at 8:00 am. Studies indicated that walls were most highly labeled after 18 hr (Mitcham et al., 1989).

Carbohydrate analyses
Total sugar content was determined using phenol-H$_2$SO$_4$ as described by Dubois et al. (1956). Uronic acids were assayed using the methods of Blumenkrantz and Asboe-Hanson (1973) or Dische (1953). Noncellulosic neutral sugars were quantified as previously described (Gross, 1984).

For glycosyl-linkage analysis, partially methylated alditol acetates were prepared according to Hakamori (1964) as described by Sanford and Conrad (1966), using modifications developed by Carpita (1984), Harris et al. (1984), and York et al. (1985). In addition, butyl-lithium, rather than potassium-methyl sulfinyl carbanion, was used (Blakeney and Stone, 1985). Polysaccharides were purified with a Sep-Pak C18 cartridge (Mort et al., 1983), and were reduced and acetylated using a procedure similar to that of Blakeney et al. (1983). Derivatives were separated and identified by relative retention time using capillary GC and subsequent spectral analysis of fragmentation patterns after electron impact mass spectrometry. They were quantified using capillary gas chromatography using a flame ionization detector according to their effective carbon response, as described by Sweet et al. (1975).

Glycosyl linkages were deduced from gas chromatography/mass spectrometry of partially methylated, partially acetylated alditol acetates. Nomenclature denotes the carbon participating in the glycosidic linkage. For example, 4-galactosyl represents a C-1 to C-4 glycosidic linkage, with the actual analyzed alditol acetate being 1,4,5-tri-O-acetyl-2,3,6-tri-O-methyl galactitol; t-arabinosyl designates non-reducing terminal arabinosyl residues.

Results and Discussion

Changes in wall composition
In an effort to determine which wall polymers are degraded and solubilized during the softening process, we have studied changes in wall carbohydrate composition during ripening. A large net decrease in tomato cell wall galactosyl content occurs during ripening (Gross, 1984; Gross and Wallner, 1979). We have shown that this loss of cell wall non-cellulosic neutral sugar residues also occurs in many other fruit types (Table 1; Gross and Sams, 1984).

Wall fractionation studies showed that the loss of galactosyl residues involves at least two separate polymers in tomato and is greatly attenuated in *rin* (ripening-inhibitor) mutant fruit (Gross, 1984). Pressey (1983) partially purified a ß-galactosidase capable of degrading a tomato cell wall galactan. However, the mechanism of cell wall galactan turnover, i.e., coordinated synthesis and degradation, remains obscure.

In a related study of wall metabolism aimed at identifying the fate of wall galactosyl residues during ripening, we observed that the content of soluble, monomeric galactose increased 5-fold during ripening; the increase did not occur in *rin* mutant fruit (Gross, 1983; Gross and Saltveit, 1982). Presumably, the increase in free, monomeric galactose occurs as a result of changes in cell wall galactan turnover (Lackey et al., 1980). The increase in free galactose is striking as this monosaccharide is generally toxic to plants (Aharoni et al., 1985; Anker, 1974; Colclasure and Yopp, 1976; Yamamoto et al., 1981).

Table 1. Changes in cell wall non-cellulosic neutral sugar during ripening of various fruit types (Gross and Sams, 1984).

Fruit type	Most abundant neutral sugar	Residue showing most significant decrease
Solanaceous fruit:		
Tomato	galactose (49%)	galactose (71%)
Bell pepper	galactose (44%)	galactose (68%)
Hot pepper	galactose (47%)	galactose (79%)
Stone fruit:		
Nectarine	arabinose (41%)	arabinose (63%)
Apricot	xylose (39%)	xylose (29%)
Plum	galactose (45%)	no change
Peach	arabinose (45%)	galactose (47%)
Cucurbit fruit:		
Cucumber	galactose (61%)	no change
Muskmelon	galactose (56%)	galactose (79%)
Melon	galactose (61%)	galactose (46%)
Squash	galactose (45%)	galactose (56%)
Small berries:		
Blueberry	xylose (55%)	arabinose (82%)
Raspberry	xylose (74%)	arabinose (33%)
Blackberry	xylose (53%)	galactose (70%)
Strawberry	xylose (29%)	arabinose (65%)
Pome fruit:		
Apple	arabinose (36%) galactose (32%)	galactose (51%)
Pear	xylose (40%) arabinose (40%)	arabinose (71%)

Stimulation of ethylene production by galactose

The toxic effect of galactose on plant cells in concert with the observed net loss of galactosyl residues from the wall and increase in soluble, monomeric galactose, led us to study the effect of exogenous galactose on unripe tomato fruit.

Infiltration of galactose into green tomato fruit resulted in a stimulation of ACC synthase activity (Fig. 1), leading to a transient increase in ACC and ethylene production and a promotion of ripening (Gross, 1985; Kim et al., 1987). Exogenous galactose also stimulated ethylene production in *rin* and *nor* tomatoes, but these fruit did not ripen; however, some red color was evident. Subsequent studies have shown that galacturonic acid, mannose, and dulcitol have similar effects (Kim et al., 1987) as galactose on "Rutgers" fruit. Clearly, further study of galactose toxicity and metabolism in relation to fruit ripening and senescence is needed.

Elicitation of ethylene by cell wall component

It is becoming increasingly clear that cell wall fragments play an important role in the regulation of plant growth and development (Albersheim and Darvill, 1985; McNeil et al., 1984). Tong et al. (1986) observed a marked increase in ethylene production by pear suspension cultures after the addition of Macerase or soluble products of Macerase-digested pear cell walls, suggesting the presence of biologically active cell wall fragments. Similar responses were found with Cellulysin in tobacco leaf discs (Anderson et al., 1982) and with tomato cell wall autolysis products (Brecht and Huber, 1986). Clearly, the relationship between the synthesis, removal, structure, and biological function of cell wall polysaccharides in relation to food quality must be addressed.

We have isolated a component from a cell wall pectic fraction that stimulates ethylene production when infiltrated into unripe fruit (Tong and Gross, 1988b). A number of cell wall fractions were infiltrated into MG tomatoes, including: IAP, CBP, and HF from fruit at various stages of ripeness. However, only CBP from MG2-MG4 tomatoes (10 µg/gfw) was found to elicit ethylene production (Fig. 2). Infiltration of CBP from MG *rin* mutant fruit into MG "Rutgers" fruit did not stimulate ethylene production. However, infiltration of "Rutgers" CBP into *rin* fruit did, suggesting that *rin* has the ability to respond to the elicitor, but does not produce it. Vertical stripes of red pigmentation developed in *rin* fruit after CBP infiltration.

The active elicitor fraction has been partially purified using DEAE-Sephadex, CM-Sephadex, and Bio-Gel P-100 column chromatography and contains protein and carbohydrate. However, the exact composition of the elicitor is unknown, pending complete purification.

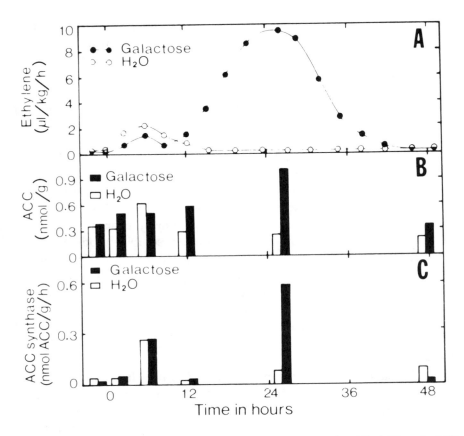

Figure 1. Effect of exogenous galactose on ethylene production (A), ACC content (B) and ACC synthase activity (C) of mature green "Rutgers" tomato fruit. Fruit were vacuum infiltrated through the stem scar with 1 ml of ddH_2O or water containing 400 μg galactose per gfw of fruit. Time 0 refers to the time of infiltration. Data are the average of three fruit. Reproduced with permission from Kim et al., 1987. Copyright 1987, The American Society of Plant Physiologists.

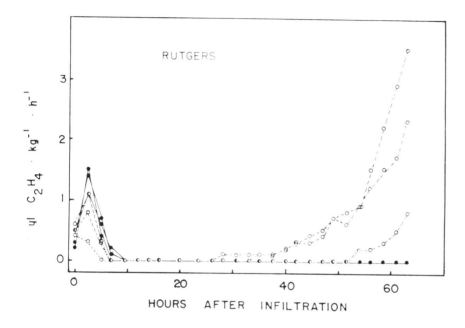

Figure 2. Ethylene production from mature green tomato fruit infiltrated with ddH$_2$O (half-closed circles), 5 mM Hepes-NaOH, pH 7 (closed circles), or CBP (open circles) from mature green "Rutgers" fruit. One µl per gfw of fruit was vacuum infiltrated through the stem scar. Ethylene was monitored by gas chromatography using an automatic, flow-through system.

Changes in wall synthesis during ripening

Although pectin degradation has been well documented, little is known about its synthesis. For a number of years, the cell wall was viewed as a static "brick wall" surrounding the cell whose function was to provide support for the plasma membrane and cell contents. Tissue textural changes were thought to occur as necessary hydrolases, such as PG, were synthesized and deposited in the cell wall where degradation could occur. This view was partly due to a lack of basic information on the structure and biosynthesis of complex carbohydrates. However, recent studies have shown that the cell wall is an extremely complex, dynamic structure. Particularly striking are the observations of Mitcham et al. (1989) on the *in vivo* synthesis of tomato fruit cell wall components during ripening. It seemed reasonable to suspect that incorporation of ^{14}C from ^{14}C-sucrose, which was injected into the pedicel of fruit on the vine, might decrease during fruit ripening. However, this was not the case (Fig. 3). Labeling of pectic, hemicellulosic, and cellulosic wall fractions occurred throughout development and ripening. In addition, there was a shift in incorporation into the various types of wall fractions synthesized after tomato fruit reached the mature-green, unripe stage. We are currently studying the composition, turnover, and structure of the various polymers synthesized during development and senescence to ascertain if a modified type of polymer is synthesized during ripening. Either *de novo* synthesis of a modified type of cell wall polymer(s), or modification of an existing polymer(s), could lead to softening of the tissue during ripening as opposed to the classically accepted dogma of a general degradation of existing polymers.

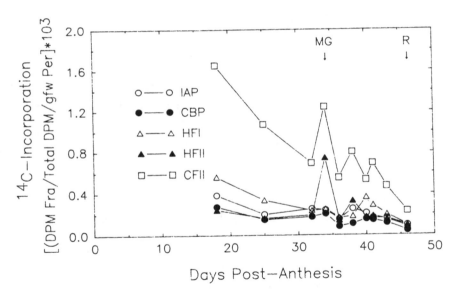

Figure 3. Incorporation of ^{14}C from $[^{14}C(U)]$-sucrose into tomato fruit cell wall fractions. Cell walls from 3 fruit at each developmental stage were combined and fractionated. Reproduced from Mitcham et al., 1989. Copyright 1989, The American Society of Plant Physiologists.

Qualitative changes in cell wall biosynthesis were also evident in studies of the glycosyl linkages present in tomato fruit cell wall hemicellulosic fractions. Huber (1983a) had reported that the molecular weight distribution of hemicelluloses changes markedly during tomato fruit ripening. We have further studied the hemicelluloses involved in this molecular weight shift using methylation analysis.

Analyses of the two hemicellulosic fractions separated by Sephacryl S-200 chromatography revealed compositional changes during ripening (Table 2). In HF II peak 1 (relatively high molecular weight), an increase in 4-xylosyl and 4-mannosyl and decreases in t-glucosyl and 4-galactosyl residues occurred during ripening. A decrease in 5-arabinosyl and increases in 4-mannosyl, 4,6-mannosyl, and 4-glucosyl residues was detected in HF II peak 2 (relatively low molecular weight). These data suggest that arabinosyl side chains are replaced by mannosyl residues and/or that synthesis of glucomannans increases during ripening. These data suggest the types of polymers that may be involved in the molecular weight shift in hemicellulose observed by Huber (1983a).

In order to understand the cell wall modifications that lead to tissue softening, it is imperative to study wall turnover. Turnover involves both degradative and synthetic reactions. The necessity to know how the catabolic enzymes are regulated *in vivo* has been recognized (Brady et al., 1987). Our results point to the need to understand mechanisms of wall polysaccharide synthesis as well. Changes in the types of polymers being incorporated into the wall at various developmental stages could have a dramatic impact on cell wall solubility and thus on cellular regulation of fruit textural quality.

Acknowledgements

The author acknowledges Jongkee Kim, Elizabeth Mitcham, and Cindy Tong for conducting and/or participating in much of the research reported herein. The author also expresses appreciation to Norman Livsey for dedicated technical support throughout the research reported in this paper. The non-ripening mutant tomato seeds were kindly donated by E.C. Tigchelaar.

Use of a company or product name by the United States Department of Agriculture does not imply approval or recommendation of the product to the exclusion of others which may also be suitable.

Table 2. Glycosyl linkage composition of HF II peak 1 and 2. The amounts of O-methylated sugars are listed as mol %. Reproduced from Tong and Gross, 1988a. Copyright 1988, Physiologia Plantarum.

Glycosyl linkage	HF II - Peak 1			HF II - Peak 2		
	MG	T	R	MG	T	R
Arabinosyl						
t-	4.3	4.2	3.6	1.1	1.3	0.7
5-	6.4	35.8	21.8	19.7	9.9	4.4
3.5-	1.1	1.1	1.0	0.7	0.6	0.1
Xylosyl						
t-	8.4	8.8	10.1	1.7	2.0	2.1
2-	5.3	5.0	3.2	0.1	0.5	0.5
4-	3.6	6.4	12.1	7.3	6.3	5.0
2,4-	2.4	3.3	3.8	1.2	1.1	0.6
Mannosyl						
t-	0.4	0.2	0.9	1.2	1.4	1.6
4-	6.1	10.4	13.2	20.9	23.8	30.4
4,6-	—	—	—	6.0	7.8	10.8
Glucosyl						
t-	2.8	0.2	0.6	1.8	2.0	0.9
4-	17.5	10.3	14.4	25.3	30.6	34.3
6-	4.9	1.5	3.1	2.3	0.8	1.3
4,6-	11.4	8.9	7.5	2.7	3.3	2.8
Galactosyl						
t-	4.1	3.8	4.7	6.9	6.0	1.9
2-	—	—	—	1.1	1.8	2.4
4-	1.1	—	—	—	—	—
3,4-	—	—	—	—	0.7	0.4

Abbreviations

ACC, 1-aminocyclopropane-1-carboxylic acid; CDTA, 1,2-diamino-cyclohexane-N,N,N',N'-tetraacetic acid; ddH_2O, distilled, deionized water; PG, polygalacturonase.

References

Aharoni, N., Philosoph-Hadas, S. and Meir, S. 1985. Carbohydrates stimulate ethylene production in tobacco leaf discs. I. Characterization of the system in relation to IAA-induced ethylene. Plant Physiol. 78: 131.

Albersheim, P. and Darvill, A.G. 1985. Oligosaccharins. Sci. Amer. 253: 58.

Anderson, J., Mattoo, A. and Lieberman, M. 1982. Induction of ethylene biosynthesis in tobacco leaf discs by cell wall digesting enzymes. Biochem. Biophys. Res. Commun. 107: 588.

Anker, L. 1974. Auxin synthesis inhibition by sugars, notably by galactose. Acta. Bot. Neerl. 23: 705.

Bartley, I.M. 1978. Exo-polygalacturonase of apple. Phytochemistry 17: 213.

Bartley, I.M. and Knee, M. 1982. The chemistry of textural changes in fruit during storage. Food Chem. 9: 47.

Blakeney, A.B. and Stone, B.A. 1985. Methylation of carbohydrates with lithium methylsuphinyl carbanion. Carbohydr. Res. 140: 319.

Blakeney, A.B., Harris, P.J., Henry, R.J. and Stone, B.A. 1983. A simple and rapid preparation of alditol acetates for monosaccharide analysis. Carbohydr. Res. 113: 291.

Blumenkrantz, N. and Asboe-Hanson, G. 1973. New method for quantitative determination of uronic acids. Anal. Biochem. 54: 484.

Brady, C.J. 1987. Fruit ripening. Ann. Rev. Plant Physiol. 38: 155.

Brady, C.J., McGlasson, W.B., Pearson, J.A., Meldrum, S.K. and Kopeliovitch, E. 1985. Interactions between the amount and molecular forms of polygalacturonase, calcium, and firmness in tomato fruit. J. Amer. Soc. Hort. Sci. 110: 254.

Brady, C.J., McGlasson, B. and Speirs, J. 1987. The biochemistry of fruit ripening. Tomato Biotechnology, pp. 279, Alan R. Liss, Inc.

Brecht, J.K. and Huber, D.J. 1986. Stimulation of tomato fruit ripening by tomato cell wall fragments. HortScience 21: 319 (Abst).

Carpita, N.C. 1984. Cell wall development in maize coleoptiles. Plant Physiol. 76: 205.

Colclasure, G.C. and Yopp, J.H. 1976. Galactose-induced ethylene evolution in mung bean hypocotyls: A possible mechanism for galactose retardation of plant growth. Physiol. Plant. 37:298.

Darvill, A., McNeil, M., Albersheim, P. and Delmer, D.P. 1980. The primary cell walls of flowering plants. In: The Biochemistry of Plants. Vol. 1, Acad. Press., p. 91.

DellaPenna, D., Alexander, D.G. and Bennett, A.B. 1986. Molecular cloning of tomato fruit polygalacturonase: analysis of polygalacturonase mRNA levels during ripening. Proc. Natl. Acad. Sci. USA. 83:6420.

Dey, P.M. and Brinson, K. 1984. Plant cell walls. In: Adv. Carbohydr. Chem. Biochem. 42:265.

Dische, Z. 1953. Qualitative and quantitative colorimetric determination of heptoses. J. Biol. Chem. 204:983.

Dubois, M., Gilles, K.A., Hamilton, J.K., Rebers, P.A. and Smith, F. 1956. Colorimetric method for determination of sugars and related substances. Anal. Chem. 28:350.

Fishman, M.L., Pepper, L., Damert, W.C., Phillips, J.G. and Barford, R.A. 1986. A critical reexamination of molecular weight and dimensions for citrus pectins. In: Chemistry and Function of Pectins. M.L. Fishman and J.J. Jen, eds. ACS Symp. Series 310.

Gross, K.C. 1983. Changes in free galactose, *myo*-inositol and other monosaccharides in normal and non-ripening mutant tomatoes. Phytochemistry 22:1137.

Gross, K.C. 1984. Fractionation and partial characterization of cell walls from normal and non-ripening mutant tomato fruit. Physiol. Plant. 62:25.

Gross, K.C. 1985. Promotion of ethylene evolution and ripening of tomato fruit by galactose. Plant Physiol. 79:306.

Gross, K.C. and Saltveit, M.E. 1982. Galactose concentration and metabolism in pericarp tissue from normal ripening and non-ripening mutant tomato fruit. J. Amer. Soc. Hort. Sci. 107:328.

Gross, K.C. and Sams, C.E. 1984. Changes in cell wall neutral sugar composition during fruit ripening: A species survey. Phytochemistry 23:2457.

Gross, K.C. and Wallner, S.J. 1979. Degradation of cell wall polysaccharides during tomato fruit ripening. Plant Physiol. 63:117.

Hakamori, S. 1964. A rapid permethylation of glycolipid and polysaccharide catalyzed by methylsulfinyl carbanion in dimethyl sulfoxide. J. Biochem. (Tokyo) 55:205.

Harris, P.J., Henry, R., Blakeney, A.B. and Stone, B.A. 1984. An improved procedure for the methylation analysis of oligosaccharides and polysaccharides. Carbohydr. Res. 127:59.

Hall, C.B. 1987. Firmness of tomato fruit tissues according to cultivar and ripeness. J. Amer. Soc. Hort. Sci. 112:663.

Hobson, G.E. 1964. Polygalacturonase in normal and abnormal tomato fruit. Biochem. J. 92:324.

Huber, D.J. 1983a. Polyuronide degradation and hemicellulose modifications in ripening tomato fruit. J. Amer. Soc. Hort. Sci. 108:405.

Huber, D.J. 1983b. The role of cell wall hydrolases in fruit softening. Hortic. Rev. 5:169.

Huber, D.J. 1984. Strawberry fruit softening: The potential roles of polyuronides and hemicelluloses. J. Food Sci. 49:1310.

Jarret, R.L., Sayama, H. and Tigchelaar, E.C. 1984. Pleiotropic effects associated with the chlorophyll intensifier mutations *high pigment* and *dark green* in tomato. J. Amer. Soc. Hort. Sci. 109:873.

Kim, J., Gross, K.C. and Solomos, T. 1987. Characterization of the stimulation of ethylene production by galactose in tomato (*Lycopersicon esculentum* Mill.) fruit. Plant Physiol. 85:804.

Labavitch, J.M. 1981. Cell wall turnover in plant development. Annu. Rev. Plant Physiol. 32:385.

Lackey, G., Gross, K.C. and Wallner, S.J. 1980. Loss of tomato cell wall galactan may involve reduced rate of synthesis. Plant Physiol. 66:532.

McNeil, M., Darvill, A.G., Fry, S.C. and Albersheim, P. 1984. Structure and function of the primary cell walls of plants. Ann. Rev. Biochem. 53:625.

Mitcham, E., Gross, K.C. and Ng, T.J. 1989. Tomato fruit cell wall synthesis during development and senescence. *In vivo* radiolabeling of wall fractions using ^{14}C-sucrose. Plant Physiol. (in press).

Mort, A.J., Parker, S. and Kuo, M-S. 1983. Recovery of methylated saccharides from methylation reaction mixtures using Sep-Pak C_{18} cartridges. Anal. Biochem. 133:380.

Pressey, R. 1983. ß-Galactosidases in ripening tomatoes. Plant Physiol. 71:132.

Sanford, P.A. and Conrad, H.E. 1966. The structure of the *Aerobacter aerogenes* A3 (S) polysaccharide. I. A reexamination using improved procedures for methylation analysis. Biochemistry 5:1508.

Selvendran, R.R. and O'Neill, M.A. 1987. Isolation and analysis of cell walls from plant material. In: Meth. Biochem. Anal. 32:25.

Sweet, D.P., Shapiro, R.H. and Albersheim, P. 1975. Quantitative analysis by various g.l.c. response-factor theories for partially methylated and partially ethylated alditol acetates. Carbohydr. Res. 40:217.

Tong, C.B. and Gross, K.C. 1988a. Glycosyl-linkage analysis of tomato fruit cell wall hemicellulosic fractions during ripening. Physiol. Plant. (In Press).

Tong, C.B. and Gross, K.C. 1988b. Elicitation of ethylene production from tomato fruit by pericarp cell wall pectic polysaccharides. Plant Physiol. 86:97 (Abstr).

Tong, C.B. and Gross, K.C. 1988c. Ripening characteristics of a tomato mutant, *dark green*, and its parent, "Manapal." (Unpublished data).

Tong, C.B., Labavitch, J.M. and Yang, S.F. 1986. The induction of ethylene production from pear cell culture by cell wall fragments. Plant Physiol. 81:929.

Yamamoto, R., Sakurai, N. and Masuda, Y. 1981. Inhibition of auxin induced cell elongation by galactose. Physiol. Plant. 53:543.

York, W.S., Darvill, A.G., McNeil, M., Stevenson, T.T. and Albersheim, P. 1985. Isolation and characterization of plant cell walls and cell wall components. Methods Enzymol. 118:3.

Prospects for the Use of Genetic Engineering in the Manipulation of Ethylene Biosynthesis and Action in Higher Plants

Anthony B. Bleecker
Division of Biology 156-29
California Institute of Technology
Pasadena, CA 91125, USA

The plant hormone ethylene is involved in a number of developmental processes and responses to stress in higher plants. We are interested in understanding how ethylene biosynthesis is regulated and in the molecular mechanisms by which ethylene induces changes in plant development. Recent experimental evidence indicates that increases in ethylene synthesis are due to the *de novo* synthesis of the rate limiting enzyme, 1-aminocyclopropane-1-carboxylic acid (ACC) synthase. The enzyme has been purified from tomato fruit tissue. Very little is known about how tissues perceive ethylene and how this signal is transduced to yield the variety of well-characterized responses. In order to gain further insight into the mechanisms of ethylene action, mutants have been isolated in the cruciferous plant *Arabidopsis thaliana* which show altered responses to ethylene. One dominant mutation, designated *etr*, shows complete insensitivity to applied ethylene in terms of a number of ethylene-mediated responses found in wild-type plants. Methodologies being used to clone genes which are involved in ethylene synthesis and action are described. The potential use of these clones in the manipulation of ethylene-mediated processes in agriculturally important plants is discussed.

Introduction

Ethylene is an endogenous regulator of growth and development in higher plants. This gaseous compound plays a number of important roles in determining the yield and quality of crop plants. In the case of climacteric fruits, ethylene can have a positive impact on food quality. One of the earliest recognizable biochemical changes which occurs as a fruit begins to ripen is an increase in ethylene synthesis (Brady, 1987; Yang and Hoffman 1984). In turn, the ethylene which is produced activates the transcription of genes that code for proteins involved in the ripening process (DellaPenna et al., 1986; Grierson et al., 1986).

Ethylene may also have a negative impact on food quality and productivity of a variety of economically important plants. Ethylene synthesis is induced in plants by a number of environmental stresses (Yang and Hoffman, 1984). The resulting elevated ethylene levels may contribute to such undesirable effects as premature flower abortion, acceleration of abscission and senescence of plant organs, and reduction of total biomass. These effects may occur during the growing season and, in the case of accelerated senescence, during post-harvest transport and handling of produce (Schouten, 1985).

Given the range of important developmental and physiological processes which are regulated by ethylene, there is interest among agricultural scientists in developing methodologies which allow the manipulation of ethylene levels and the responsiveness of plant tissues to ethylene. Chemicals such as ethephon (Ables, 1973), which produce ethylene when in contact with plant tissues, are in use in a number of agricultural systems. Several inhibitors of ethylene biosynthesis are available which have been invaluable to basic research (Yang and Hoffman, 1984). However, these compounds are often nonspecific and potentially toxic and have not been exploited on a commercial level. There are also compounds available which block ethylene action in plants, most notably the competitive inhibitor, 2,5-norbornediene (Sisler and Yang, 1984). Unfortunately, these compounds generate the same problem of potential toxicity and so have not proven useful to agriculture. Mechanical methods for reducing the levels of ethylene during post-harvest transport are widely used commercially, but are only moderately effective in reducing endogenous ethylene levels (Knee, 1985).

As we gain a better understanding at the biochemical level of how ethylene synthesis is regulated and of how plant tissues perceive ethylene and transduce this signal into a variety of responses, we may begin to consider the possible ways in which genetic engineering may allow us to alter ethylene regulatory pathways in agriculturally significant ways. In this chapter, I will review some recent findings on ethylene synthesis and action and consider how this information may provide us with new strategies for improving the yield and quality of agriculturally important plant species.

Regulation of Ethylene Synthesis

In all higher plants investigated, ethylene is synthesized via a single pathway (Yang and Hoffman, 1984). S-adenosyl-L-methionine (SAM) is converted to the intermediate 1-aminocyclopropane-1-carboxylic acid (ACC) by the enzyme ACC synthase. The ACC is subsequently oxidized to ethylene by the ethylene-forming enzyme (EFE). The conversion of SAM to ACC appears to be the primary rate limiting step in ethylene synthesis. A good correlation exists in a number of systems between the rate of ethylene synthesis and the amount of extractable ACC synthase activity. Furthermore, a number of environmental conditions which elicit an increase in ethylene synthesis have been shown to induce an increase in ACC synthase activity. Work with protein synthesis inhibitors indicated that the increased ACC synthase activity was due to *de novo* formation of the ACC synthase protein (Acaster and Kende, 1983).

The ACC synthase enzyme was first purified to homogeneity from tomato fruit tissue (Bleecker et al., 1986). Recent reports indicate that this enzyme has also been purified from mung bean and squash (Tsai et al., 1988, Nakajima et al., 1988). Monoclonal antibodies (MAbs) directed against the tomato enzyme were used to demonstrate that wound induced increases in the level of ACC synthase activity in the tomato fruit resulted from *de novo* synthesis of the 50 kDa ACC synthase protein (Bleecker et al., 1988). These results are consistent with the idea that increases in enzyme activity were due to transcriptional activation of the ACC synthase gene. More specific information regarding the mechanism by which ACC synthase levels are regulated in plant tissues awaits the cloning of the gene.

Mode of Action of Ethylene

Investigation of the molecular mechanisms by which ethylene action is mediated requires a different approach since we know almost nothing about the biochemical pathways involved in ethylene responses. For this reason we undertook a genetic approach using the small curcifer *Arabidopsis thaliana*. Using the ethylene-mediated inhibition of seedling growth as an assay, mutants with altered responses to ethylene were isolated from a chemically mutagenized population of seed. One mutant, designated *etr*, has been physiologically and genetically characterized (Bleecker et al., 1988b). In addition to showing complete insensitivity to ethylene-mediated inhibition of hypocotyl and root elongation, the dominant *etr* mutant also lacks a number of other ethylene responses which occur in wild-type plants. These include promotion of seed germination, acceleration of leaf senescence, enhancement of peroxidase activity, and feedback inhibition of ethylene synthesis.

The fact that a single mutation affects a number of ethylene responses occurring in different tissues and at different developmental stages indicates that a single receptor system is responsible for all of these responses in *Arabidopsis*. Measurements of saturable ethylene binding in intact leaf tissue showed a 5-fold reduction in binding in the *etr* mutant relative to wild-type tissue (Bleecker et al., 1988b) indicating that the *etr* mutation may directly affect the receptor for ethylene.

Since the *etr* locus may code for a gene product which is involved in ethylene receptor function, a good deal of information could be gained about the biochemical nature of the receptor if a clone for the *etr* gene could be obtained. Efforts are currently underway to clone the *etr* gene using the technique of chromosome walking (Meyerowitz, 1987). This strategy involves placing the *etr* mutation on a restriction fragment-length polymorphism (RFLP) linkage map that is composed of markers consisting of randomly cloned sequences of *Arabidopsis* genomic DNA (Chang et al., 1988). The RFLP marker which maps closest to the mutation is then used to isolate additional overlapping clones from a genomic library. This process is repeated until a region of the chromosome which encompasses the mutation has been cloned. The cloned sequences are then transferred back into *Arabidopsis* plants via *Agrobacterium*-mediated transformation and the sequences containing the *etr* mutant locus are identified by their ability to confer ethylene insensitivity in transgenic plants.

Implication for the Future

With the advent of technologies which allow for the analysis and transfer of genetic material in an ever-increasing number of agriculturally important crop species, it is of interest to consider the possible ways in which the work in progress outlined above may be used to alter the ethylene regulatory pathways in plants in commercially favorable ways. The cloning of the gene which codes for ACC synthase could lead to a number of strategies for engineered regulation of ethylene biosynthetic rates. Strategies for producing increases in ethylene synthesis could take advantage of an increasing variety of characterized temporal and tissue-specific transcriptional regulatory elements that operate in plants. These elements could be fused to the ACC-synthase protein-coding sequences and the resulting chimeric genes transferred into the plant of interest. These constructs would direct tissue- and/or stage-specific increases in ethylene synthesis in transgenic plants. Alternatively, ethylene synthesis could be inhibited through the construction of an antisense ACC synthase gene which, when expressed, would produce an RNA which would hybridize to and thus inactivate the ACC synthase message which is normally produced. Both of these techniques have been used successfully with other plant genes (Bennett, this volume; Smith et al., 1988).

The cloning of genes involved in ethylene responses could allow for the genetic manipulation of the sensitivity of plant tissues to ethylene. For example, the dominant nature of the *etr* mutation opens the possibility for reducing or eliminating ethylene responses in a stage- or tissue-specific way using chimeric constructs which express the mutant gene. In addition, an increase in ethylene sensitivity could potentially be produced by over-expression of the wild-type gene.

While these possibilities are, at present, speculative, they could potentially lead to a number of strategies for improving food quality. For example, inhibition of ethylene synthesis in climacteric fruit would prevent or delay many aspects of ripening. Harvested fruit could subsequently be induced to ripen by ethylene treatment after the fruit arrives at its final destination. On the other hand, inhibition of ethylene action could increase the shelf life of a number of fruits and vegetables which are normally subject to ethylene-mediated acceleration of senescence.

Acknowledgements

The work on ethylene mutants in *Arabidopsis* has been supported by grants from the U.S. Department of Energy to Hans Kende (contract DE-ACO2-79ER01338) and to Elliot Meyerowitz (grant DE-FG03-88ER13873).

References

Ables, F.B. 1973. Ethylene in Plant Biology. Academic Press, New York.

Acaster, M.A. and Kende, H. 1983. Properties and partial purification of 1-aminocyclopropane-1 caboxylate synthase. Plant Physiol. 72, 139-145.

Bleecker A.B., Kenyon, W.H., Somerville, S.C. and Kende, H. 1986. Use of monoclonal antibodies in the purification and characterization of 1-aminocyclopropane-1-carboxylate synthase, an enzyme in ethylene biosynthesis. Proc. Nat. Acad. Sci. 83, 7755-7759.

Bleecker, A.B., Robinson, G. and Kende, H. 1988. Studies on the regulation of 1-aminocyclopropane-1-carboxylate synthase in tomato using monoclonal antibodies. Planta 173: 385-390.

Bleecker, A.B., Estelle, M.A., Somerville, C. and Kende, H. 1988. Insensitivity to ethylene confered by a dominant mutation in *Arabidopsis thaliana*. Science 241: 1086-1089.

Brady, C.J. 1987. Fruit ripening. Ann. Rev. Plant. Physiol. 38: 155-178.

Chang, C., Bowman, J.L., DeJohn, A.W., Lander, E.S. and Meyerowitz, E.M. 1988. Restriction fragment length polymorphism linkage map for *Arabidopsis thaliana*.

DellaPenna, D., Alexander, D.C. and Bennett, A.B. 1986. Molecular cloning of tomato fruit polygalacturonase: analysis of polygalacturonase mRNA levels during ripening. Proc. Nat. Acad. Sci. 83: 6420-6424.

Grierson, D., Maunders, M.J., Slater, A., Ray, J., Bird, C.R., Schuch, W., Holdsworth, M.J., Tucker, G.A. and Knapp, J.E. 1986. Gene expression during tomato ripening. Phil. Trans. R. Soc. Lond. 314: 399-410.

Knee, M. 1985. Evaluating the practical significance of ethylene in fruit storage. In: Ethylene in Plant Development, Roberts, J.A. and Tucker, G.A. (eds.), Butterworths, London, pp. 297-302.

Meyerowitz, E.M. 1987. *Arabidopsis thaliana*. Ann. Rev. Genet. 21: 93-111.

Nakajima., N. Nakagawa, N. and Imaseki, H. 1988. Molecular size of wound induced 1-aminocyclopropane-1-carboxylate synthase from *Cucurbita maxima* and changes of translatable mRNA of the enzyme after wounding. Plant Cell Physiol. 29: 989-998.

Schouten, S.P. 1985. Significance of ethylene in post-harvest handling of vegetables. In: Ethylene in Plant Development, Roberts, J.A. and Tucker, G.A. (eds.), Butterworths, London, pp. 356-362.

Sisler, E.C. and Yang, S.F. 1984. Anti-ethylene effects of cis-2-butene and cyclic olefins. Phytochemistry 23: 2765-2768.

Smith, C.J.S., Watson, C.F., Ray, J., Bird, C.R., Morris, P.C., Schuch, W. and Grierson, D. 1988. Antisense RNA inhibition of polygalacturonase gene expression in transgenic plants. Nature 334: 724-726.

Tsai, D.S., Arteca, R.N., Bachman, J.M. and Phillips, A.T. 1988. Purification and characterization of 1-aminocyclopropane-1-carboxylate synthase from etiolated mung bean hypocotyls. Arch. Biochem. Biophy. 264: 632-640.

Yang, S.F. and Hoffman, N.T. 1984. Ethylene biosynthesis and its regulation in higher plants. Ann. Rev. Plant Physiol. 35: 155-190.

Yoshii, H. and Imaseki, H. 1982. Regulation of auxin induced ethylene biosynthesis. Plant and Cell Physiol. 23: 639-649.

Tomato Fruit Polygalacturonase: Gene Regulation and Enzyme Function

Alan B. Bennett and Dean DellaPenna
Mann Laboratory, Department of Vegetable Crops
University of California
Davis, CA 95616, USA

Robert L. Fischer, James Giovannoni, and James E. Lincoln
Division of Molecular Plant Biology
University of California
Berkeley, CA 94720, USA

Polygalacturonase has been implicated as the primary enzymatic determinant of pectin degradation that accompanies tomato fruit ripening. This pectin degradation, in turn, has been proposed to be the major structural change in the cell wall that contributes to tomato fruit softening. Consequently, polygalacturonase has been targeted for molecular genetic manipulation with the aim of improving tomato fruit quality by altering the timing and/or extent of fruit softening. However, the actual contribution of polygalacturonase in fruit softening has been drawn largely from circumstantial evidence. In order to critically assess polygalacturonase as a target for molecular genetic manipulation, we have characterized regulation of the polygalacturonase gene and used this information to devise and implement a molecular genetic strategy to manipulate levels of polygalacturonase expression. These experiments are allowing us to critically define the function(s) of polygalacturonase in ripening tomato fruit.

Introduction

Ripening represents a major transition in tomato fruit development. In addition to visually obvious changes, such as color development, major changes in cell wall structure accompany tomato fruit ripening. The cell wall is comprised primarily of cellulose, hemicellulose, and pectins and provides structural rigidity to the fruit tissue. Changes in cell wall structure that accompany fruit ripening are thought to be responsible for the textural changes and softening associated with ripening. The quality of fresh market tomatoes is closely linked to the softening process. In order to prevent damage during harvest, packing, and shipping, fresh market tomatoes are harvested at a firm, unripe stage, prior to the development of full flavor. A delay in softening should therefore allow for greater maturation of the fruit on the vine prior to harvest. Pectins contribute to the quality of processed tomatoes. Altering the extent of pectin degradation that naturally occurs during ripening and in the early stages of processing may also be of some benefit in improving the quality of processed tomatoes.

Although numerous subtle changes in cell wall structure occur during fruit ripening, a major structural change is the degradation of pectins or polyuronides (Wallner and Bloom, 1977). While polyuronide degradation is best examined biochemically, it can also be observed as a loss of the darkly staining middle lamella in electron micrographs (Crookes and Grierson, 1983). A single enzyme, polygalacturonase (PG), has been implicated as the sole agent of polyuronide degradation in ripening tomato fruit (Huber, 1981; Themmen et al., 1982; Wallner and Bloom, 1977). In addition, PG-dependent polyuronide degradation has been suggested to be the major determinant of fruit softening (Huber, 1983; Hobson, 1965). It has also been speculated that PG-dependent polyuronide degradation may release biologically active pectic fragments that regulate other components of the ripening process (Baldwin and Pressey, 1988; Bennett and DellaPenna, 1987B; Brady, 1987). Together, these results implicate PG as a key enzyme in polyuronide degradation and fruit softening and perhaps in regulating or coordinating other components of the ripening process.

If these proposed roles of PG are in fact true, then the importance of PG in contributing to a major component of fruit quality (i.e., softening) would be surpassed by its role as an endogenous regulator of the ripening process. In the context of designing strategies to improve tomato fruit quality through the molecular genetic manipulation of PG expression it is first important to clearly understand the role of PG in the ripening process. This will in turn dictate: 1) whether PG should be manipulated and 2) whether PG can be manipulated independently of the ripening process or whether the two events (i.e., softening and ripening) are inextricably linked. Because of a level of ambiguity inherent in all of the experiments to date analyzing PG function, the actual role of PG in softening and other aspects of ripening has remained somewhat controversial. The goal of our research has been to resolve this controversy by first understanding the regulation of PG gene expression and then using this information to implement molecular genetic strategies designed to critically assess the physiological function of PG in the ripening process. In this paper, I will review our progress in this endeavor.

Regulation of PG Gene Expression

It has been proposed, and a great deal of supporting evidence has been presented, indicating that ripening-associated alterations in metabolism result from specific alterations in gene expression (Biggs et al., 1986; DellaPenna et al., 1986; Lincoln et al., 1987; Mansson et al., 1985; Maunders et al., 1987; Rattanapanone et al., 1978; Speirs et al., 1984). Presumably, alterations in cell wall metabolism could be brought about, similarly, by alterations in the expression of genes encoding cell wall enzymes. Although the expression of many ripening-associated genes of unknown function have been studied, PG is the only ripening-associated gene of known function whose expression has been extensively characterized.

PG was first characterized in tomato fruit in 1964 (Hobson) and its activity shown to increase with ripening of the fruit. Subsequent studies indicated that the increase in PG activity results from *de novo* synthesis of the protein (Brady et al., 1982; Tucker and Grierson, 1982). In order to further examine the regulation of PG, a PG cDNA clone was identified by antibody screening of a ripe tomato fruit expression library, and the PG cDNA clone used to directly examine PG mRNA levels during fruit ripening (DellaPenna et al., 1986). The results of these experiments indicated that PG mRNA was undetectable in mature green tomato fruit, appeared early in fruit ripening, and continued to accumulate throughout ripening (Fig. 1). We later showed that PG mRNA is quite abundant, accumulating to 1.2% of the poly $(A)^+$ RNA mass in ripe fruit of the variety cv. Castle-mart (Bennett and DellaPenna, 1987A). In subsequent experiments, we have observed levels of PG mRNA exceeding 4.6% of the poly $(A)^+$ RNA mass in ripe fruit of a different variety (cv. Rutgers).

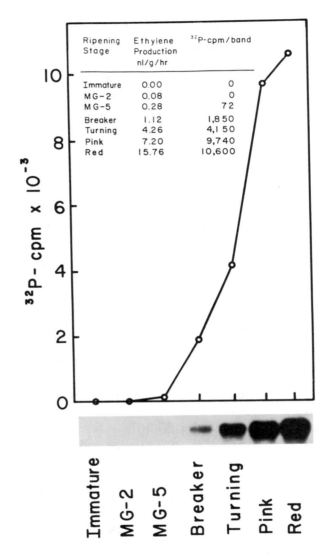

Figure 1. RNA gel blot analysis of poly(A)$^+$ RNA isolated from fruit of the indicated ripening stage, with MG-2 and MG-5 referring to early and late stages of mature green fruit, respectively. One υg of RNA was separated by agarose gel electrophoresis, blotted to nitrocellulose and probed with the ^{32}P-labeled PG cDNA, pPG16. The autoradiogram (lower panel) indicated a single size class of PG mRNA (approximately 1.8 kb) that first became detectable (upon prolonged exposure) at the MG-5 stage. Each band was excised and radioactivity determined and plotted in the upper panel. Ethylene production at each ripening stage is indicated in the inset. (Reprinted from DellaPenna et al., 1986, with permission.)

To assess whether the accumulation of PG mRNA in ripening fruit resulted from increased gene transcription, we assayed transcriptional activity of the PG gene by nuclear run-on transcription. Our results have demonstrated that transcription of the PG gene is undetectable in mature green fruit, becomes detectable at the mature green-3 stage of ripening and the gene remains transcriptionally active in ripe fruit (DellaPenna, Lincoln, Fischer and Bennett, unpublished). In addition, transcriptional activity of the PG gene was repressed in the ripening-impaired mutants, *rin*, *nor* and *Nr*. Together these results indicate that transcription of the PG gene is an important control point regulating PG expression in ripening tomato and that PG expression in the ripening-impaired mutants is blocked at the transcriptional level.

In the *in vitro* transcription assays, the transcriptional activity of other genes was assessed for comparison to PG gene transcription. Interestingly, in ripe fruit the rate of PG gene transcription was quite low in relation to its mRNA abundance as compared to the other genes examined. This observation suggests that, while developmental timing of PG expression is transcriptionally regulated, post-transcriptional processes, such as high mRNA stability, may contribute to the high levels of PG mRNA accumulation observed in ripe fruit.

Maturation of the PG Protein

In addition to processes regulating PG mRNA levels in ripening fruit, we have begun to examine potential sites of post-translational regulation of PG. This possible level of regulation is suggested: 1) by of the existence of three isozymes of PG that may differ in their function in cell wall degradation and 2) by the fact that PG is a secreted protein, therefore undergoing both co-translational and post-translational processing associated with translocation through the endomembrane system.

Because antibodies to one PG isozyme (2A) cross-react with the other two isozymes (1 and 2B) (Ali and Brady, 1982) and because a single mRNA species appears to give rise to the three PG isozymes (DellaPenna et al., 1986), it has been proposed that the catalytic subunits of all three isozymes may be derived from a single gene. We provided further support for this by showing that PG 2A and 2B differ from one another by their degree of glycosylation, which apparently results from heterogeneous glycosylation upon entry of the nascent polypeptide into the endoplasmic reticulum (DellaPenna and Bennett, 1988). Conclusive evidence that PG I, 2A, and 2B are derived from one gene has now been obtained by forcing expression of a single PG gene in transgenic *rin* tomato fruit and observing the production of all three PG isozymes (Giovannoni, DellaPenna, Bennett and Fischer, unpublished; see below).

Early studies of *in vitro* translation of PG mRNA indicated that the primary PG translation product was 54kD as compared to the size of the mature PG2 protein of 45/46 kD (Sato et al., 1984; DellaPenna et al., 1986). Isolation and sequencing of full-length cDNA clones indicated that the size discrepancy was due to a 71 amino acid N-terminal extension (Bennett and DellaPenna, 1987A; DellaPenna and Bennett, 1988; Grierson et al., 1986; Sheehy et al., 1987), and a 13 amino acid C-terminal extension (Sheehy et al., 1987) both of which are encoded by the cDNA but not found on the mature protein. We carried out an analysis of *in vitro* processing of the primary PG *in vitro* translation product to determine if the 71 amino acid N-terminal extension represented single or multiple proteolytic processing domains (DellaPenna and Bennett, 1988). The results of this analysis indicated that the N-terminal extension consisted of at least two processing domains. The first is a typical hydrophobic signal sequence cleaved cotranslationally between Ser(24) and Asn(25) (Fig. 2).

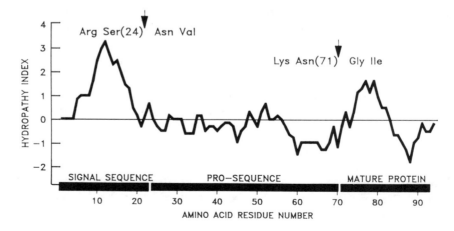

Figure 2. Hydropathy plot of the first 93 amino acids of PG, showing the relative size, cleavage sites, and hydropathy of the signal sequence and pro-peptide.

By analogy to many other proteins this domain is responsible for directing co-translational insertion of the nascent polypeptide into the endoplasmic reticulum. A second domain of 47 amino acids is apparently cleaved post-translationally after entry of the protein into the endoplasmic reticulum but prior to maturation of the protein in the cell wall. This second processing domain is cleaved between Asn(71) and Gly(72) (Fig. 2), a cleavage site similar to that observed in the post-translational proteolytic cleavage of many vacuolar proteins. The function of this second processing domain, referred to as the pro-sequence, is unknown but may be involved in maintaining the protein in an inactive state during secretion or in facilitating translocation of the protein to the cell wall. In either case processing of this domain may potentially contribute to a further post-translational level of regulation of PG expression. A model summarizing the maturation of PG is shown in Fig. 3.

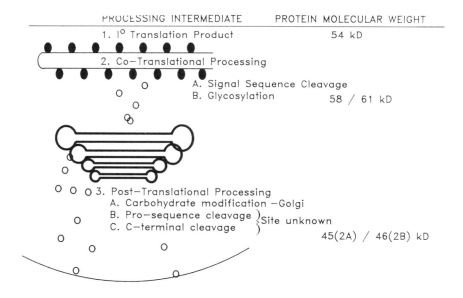

Figure 3. Diagram summarizing the steps of co-translational and post-translational processing of PG. Subcellular sites of processing are indicated in the diagram, with co-translational processing occurring in the endoplasmic reticulum and post-translational carbohydrate modification occurring in the Golgi. Sites of post-translational N-terminal and C-terminal proteolytic post-translational processing are not known.

Function of PG

As discussed above, PG has been proposed to be the primary determinant of both pectin degradation and fruit softening during tomato ripening. It has been further suggested that PG activity may release biologically active pectic fragments that contribute to activation of other components of the ripening process. To critically assess these proposed functions of PG it is necessary to design molecular genetic strategies to specifically modify PG expression and then carefully monitor the phenotypic consequences of this modified PG expression. There are basically two approaches to carrying out such an experiment:

1) The first approach forces the expression of PG in a genetic background that does not normally express PG. The phenotypic effects of expressing PG in this background can then be characterized. In tomato there are several ripening-impaired mutants (*rin*, *nor* and *Nr*) that are deficient in many ripening-associated processes, including PG expression. By forcing expression of PG in the mutant genotypes it is possible to determine what aspects of ripening (i.e., pectin degradation, fruit softening, fruit ripening) have been restored by "adding back" PG. We refer to this approach as "mutant complementation."

2) The second approach uses antisense RNA to depress PG expression in wild-type tomato fruit and examines the consequence of this reduced PG expression. This approach involves the introduction of a PG gene into tomato in its reverse orientation. When the endogenous PG gene is expressed along with the reverse orientation or antisense gene, RNA/RNA hybrids form which are inactive. In this manner it is possible to reduce the expression of an endogenous gene (in this case PG) in any genetic background. Although complete depression of expression is difficult to achieve, this approach provides a means to test the function of PG by decreasing its normal expression level and monitoring the effects on pectin degradation, fruit softening, and fruit ripening.

The two approaches provide complementary information that, together, should provide a comprehensive assessment of the function of PG.

Mutant complementation of PG

Our experiments have focused on the first approach, that of mutant complementation. We have employed the ripening-impaired mutant *rin* as the PG null genotype. In the *rin* genotype essentially all ripening-associated changes (ethylene production, lycopene synthesis, softening and PG expression) are inhibited. A ripening-associated gene (of unknown function) referred to as E8 (Lincoln et al., 1987) has been found to be coordinately regulated with PG in normally ripening fruit. We have recently observed that the rate of E8 gene transcription is approximately 7 times greater than that of the PG gene (DellaPenna, Lincoln, Fischer and Bennett, unpublished) and, fortuitously, the E8 gene is transcriptionally active in *rin* fruit whereas the PG gene is transcriptionally inactive in *rin*. This observation suggested a strategy of constructing a chimeric gene by fusing the E8 5' promoter region to the structural PG gene (Fig. 4). Our previous data suggested that such a chimeric gene should be functional in *rin*, resulting in the expression of PG in this mutant genotype at the appropriate developmental time.

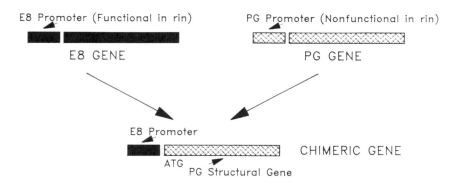

Figure 4. Diagram illustrating the construction of the chimeric gene comprised of a 5' upstream region of E8 and the coding sequence of the PG gene. This chimeric gene was transfected into *rin* tomato plants by *Agrobacterium*-mediated transformation.

The E8/PG chimeric gene was constructed and transformed into *rin* (genetic background cv. Ailsa Craig) tomato plants. Because the E8 gene is ethylene-inducible (Lincoln et al., 1987) the fruit were analyzed by harvesting after 35 days and holding in air or 500 ppm propylene. In this case propylene was used as an ethylene analogue to promote expression of the E8/PG chimeric gene so that we could simultaneously stimulate expression of the E8/PG chimeric gene and monitor ethylene production by the fruit (McMurchie et al., 1972). Protein extracted from the fruit cell wall was analyzed by SDS-PAGE and immunoblotting and it was found that after 11 days in propylene, fruit from each of three transgenic plants were expressing elevated levels of PG (Fig. 5). The proteins analyzed were isolated from a cell wall fraction and the PG present appeared to be processed to its normal size, suggesting the chimeric gene was competent in producing PG protein that was appropriately processed and targeted to the cell wall in *rin* fruit.

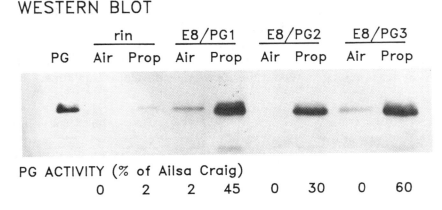

Figure 5. PG expression in *rin* fruit was analyzed by western blot analysis using antibodies raised against pure PG. Pure PG protein was run in lane 1, lanes 2-9 are cell wall protein extracts from *rin* fruit or from fruit of three different transgenic plants that contain the E8/PG chimeric gene. Fruit were harvested and held in air or 500 ppm propylene for 11 days prior to extraction. PG protein was detected in transgenic fruit (labeled E8/PG1,2,and3) if the fruit were held in 500 ppm propylene but not when held in air. The immunologically detectable protein was active when assayed *in vitro* as indicated in the lower panel.

Our analysis of the transgenic *rin* fruit is incomplete but present results indicate that:

1) Expression of PG in *rin* fruit does result in polyuronide degradation (as measured by the production of EDTA-soluble polyuronides) at levels comparable to wild-type fruit, indicating that the enzyme is active *in vivo*.

2) Expression of PG in the *rin* genotype does not result in ethylene production or the enhancement of ripening, at least as indicated by lycopene production.

At the present time we cannot conclusively state whether PG expression in the *rin* genotype has any effect on fruit softening, due to inaccuracies associated with measuring this parameter. Our results do indicate, however, that fruit softening of the transgenic *rin* fruit is far less than in wild-type fruit, suggesting that PG is not the sole determinant of fruit softening. Overall, our results indicate that PG is most likely responsible for polyuronide degradation in ripening tomato fruit. However, our results do not support a role for PG in activating other components of the ripening process and, while presently inconclusive regarding the quantitative relationship of PG to fruit softening, suggest that PG is not the sole enzymic determinant of softening.

Antisense depression of PG
Other research groups have focused their experiments on the use of antisense RNA to depress the accumulation of PG in normally ripening tomato genotypes (Smith et al., 1988; Sheehy et al., in press). Somewhat surprisingly these experiments have led to a dramatic reduction in the expression of PG despite the fact that PG mRNA is present at very high levels in ripening tomato fruit. Smith et al. (1988) have reported a 94% reduction in PG mRNA levels and a 90% reduction in PG enzyme activity in transgenic tomato fruit that carry the antisense PG construction. In these fruit ripening appeared to occur normally, as judged by lycopene accumulation, but perhaps more interesting, softening of the fruit was unaffected by the 90% reduction in PG enzyme levels (Smith et al., in press). These results suggest that high levels of PG are not required for normal fruit ripening and softening to occur.

Conclusions

Our studies of PG expression have elucidated, at least in general terms, the mechanisms by which this gene is regulated during fruit ripening and have provided additional insights into post-translational processing of cell wall proteins. These studies have also led to the design and implementation of a molecular genetic strategy to assess the physiological function of PG by forcing its expression in a ripening-impaired mutant genotype. Other groups have similarly employed molecular genetic information and tools to reduce PG expression using antisense RNA, again to critically assess the function of PG. In a general sense these experiments provide examples of the power of applying of molecular genetic approaches to address an important and persistent physiological question. More specifically, these experiments should clearly define the function of PG in tomato fruit ripening. The present results from our experiments forcing PG expression in *rin* fruit and those of Smith et al. (1988) using antisense RNA to depress PG expression are consistent with each other, in that expression of PG in *rin* fruit did not promote ripening nor did the depression of PG in wild-type fruit inhibit ripening, at least as measured by lycopene accumulation. Both experiments also failed to demonstrate a close relationship between PG levels and tomato fruit softening, suggesting that PG may not, as has been assumed for many years, be the exclusive determinant of fruit softening. Further analysis of both mutant complementation and antisense experiments should provide more comprehensive conclusions regarding the function of PG. The results of this analysis will provide a critical basis in determining the appropriateness of PG as a target for molecular genetic manipulation with the aim of improving tomato fruit quality.

Acknowledgements

This research was supported by United States Department of Agriculture-Competitive Research Grants Office grant 87-CRCR-1-2525 and by gifts from Chesebrough-Ponds and Beatrice/Hunt Wesson.

References

Ali, Z.M. and Brady, C.J. 1982. Purification and characterization of the polygalacturonases of tomato fruits. Aust. J. Plant. Physiol. 9: 155-169.

Baldwin, E.A. and Pressey, R. 1988. Tomato polygalacturonase elicits ethylene production in tomato fruit. J. Amer. Soc. Hort. Sci. 113: 92-95.

Bennett, A.B. and DellaPenna, D. 1987A. Polygalacturonase gene expression in ripening tomato fruit. In "Tomato Biotechnology," pp. 299-308. Alan R. Liss, New York.

Bennett, A.B. and DellaPenna, D. 1987B. Polygalacturonase: its importance and regulation in ripening. In "Plant Senescence: Its Biochemistry and Physiology." pp 98-107. Amer. Soc. Plant Physiol., Rockville, MD.

Biggs, M.S., Harriman, R.W. and Handa, A.K. 1986. Changes in gene expression during tomato fruit ripening. Plant Physiol. 81: 395-403.

Brady, C.J. 1987. Fruit ripening. Annu. Rev. Plant Physiol. 38: 155-178.

Brady, C.J., MacAlpine, G., McGlasson, W.B. and Ueda, Y. 1982. Polygalacturonase in tomato fruits and the induction of ripening. Aust. J. Plant Physiol. 9: 171-178.

Crookes, P.R. and Grierson, D. 1983. Ultrastructure of tomato fruit ripening and the role of polygalacturonase isoenzymes in cell wall degradation. Plant Physiol. 72: 1088-1093.

DellaPenna, D., Alexander, D.C. and Bennett, A.B. 1986. Molecular cloning of tomato fruit polygalacturonase: Analysis of polygalacturonase mRNA levels during ripening. Proc. Natl. Acad. Sci. USA 83: 6420-6424.

DellaPenna, D. and Bennett, A.B. 1988. *In vitro* synthesis and processing of tomato fruit polygalacturonase. Plant Physiol. 86: 1057-1063.

Grierson, D., Tucker, G.A., Keen, J., Ray, J., Bird, C.R. and Schuch, W. 1986. Sequencing and identification of a cDNA clone for tomato polygalacturonase. Nucl. Acids Res. 14: 8595-8603.

Hobson, G.E. 1964. Polygalacturonase in normal and abnormal tomato fruit. Biochem. J. 92: 324-332.

Hobson, G.E. 1965. The firmness of tomato fruit in relation of polygalacturonase activity. J. Hort. Sci. 40: 66-72.

Huber, D.J. 1981. Polyuronide degradation and hemicellulose modifications in ripening tomato fruit. J. Amer. Soc. Hort. Sci. 108: 405-409.

Huber, D.J. 1983. The role of cell wall hydrolases in fruit softening. Hort. Rev. 5: 169-219.

Lincoln, J.E., Cordes, S., Read, E. and Fischer, R.L. 1987. Regulation of gene expression by ethylene during tomato fruit development. Proc. Natl. Acad. Sci. USA 84: 2793-2797.

Mansson, P.E., Hsu, D. and Stalker, D. 1985. Characterization of fruit specific cDNAs from tomato. Mol. Gen. Genet. 200: 356-361.

Maunders, M.J., Holdsworth, M.J., Slater, A., Knapp, J.E., Bird, C.R., Schuch, W. and Grierson, D. 1987. Ethylene stimulates the accumulation of ripening-related mRNAs in tomatoes. Plant Cell and Environ. 10: 177-184.

McMurchie, E.J., McGlasson, W.B. and Eaks, I.L. 1972. Treatment of fruit with propylene gives information about the biogenesis of ethylene. Nature 237: 235-236.

Rattanapanone, N,, Spiers, J. and Grierson, D. 1978. Evidence for changes in messenger RNA content related to tomato fruit ripening. Phytochem. 17: 1485-1486.

Sato, T., Kusaba, S., Nakagawa, H. and Ogura, N. 1984. Cell free synthesis of a putative precursor of polygalacturonase in tomato fruits. Plant and Cell Physiol. 25(6): 1069-1071.

Sheehy, R.E., Pearson, J., Brady, C.J. and Hiatt, W.R. 1987. Molecular characterization of tomato fruit polygalacturonase. Mol. Gen. Genet. 208: 30-36.

Sheehy, R.E., Kramer, M. and Hiatt, W.R. 1988. Reduction of polygalacturonase activity in tomato fruit by antisense RNA. Proc. Natl. Acad. Sci. USA (in press).

Smith, C.J.S., Watson, C.F., Ray, J., Bird, C.R., Morris, P.C., Schuch, W. and Grierson, D. 1988. Antisense RNA inhibition of polygalacturonase gene expression in transgenic tomatoes. Nature 334: 724-726.

Speirs, J., Brady, C.J., Grierson, D. and Lee, E. 1984. Changes in ribosome organization and mRNA abundance in ripening tomato fruit. Aust. J. Plant. Physiol. 11: 225-233.

Themmen, A.P.N., Tucker, G.A. and Grierson, D. 1982. Degradation of isolated tomato cell walls by purified polygalacturonase *in vitro*. Plant Physiol. 69: 122-124.

Tucker, G.A. and Grierson, D. 1982. Synthesis of polygalacturonase during tomato fruit ripening. Planta 155: 64-67.

Wallner, S.J. and Bloom, H.L. 1977. Characteristics of tomato cell wall degradation *in vitro*. Implication for the study of fruit-softening enzymes. Plant Physiol. 60: 207-210.

Molecular Interactions of Contractile Proteins

Marion L. Greaser, Keh-Ming Pan, Jeffrey D. Fritz,
Laura J. Mundschau, and Peter H. Cooke
Department of Meat and Animal Science and Integrated Microscopy Resource
University of Wisconsin
Madison, WI 53706, USA

The contractile proteins of muscle are assembled in intricate, highly ordered structures termed myofibrils. These organelles function to produce movement during life and play an essential textural role in muscle used for food. The structure, composition, and protein-protein interactions in the myofibril will be reviewed. Special emphasis will be directed toward our recent studies on titin, a protein with a subunit size greater than 1,000,000 daltons. Monoclonal and polyclonal antibodies, in conjunction with immunofluorescence and high-voltage electron microscopy, have been used to find the regions of the myofibril in which titin is located. The titin antibody positions translocate in myofibrils subjected to sodium chloride concentrations similar to those found in meat processing. There is also a change in titin location that occurs with postmortem storage. These results suggest that the location and/or structural integrity of titin may play an important role in meat quality.

Introduction

The use of muscle as food and the application of biotechnology to improve its quality requires a detailed understanding of its structure and composition. A brief review of our current knowledge about muscle will be presented. Additional details may be obtained from the following books and reviews: Squire (1986), Ruegg (1988), Ohtsuki et al. (1986), Leavis and Gergely (1984), Greaser (1986), and King and Macfarlane (1987).

Muscle Structure

Since the classical studies of Albert Szent-Gyorgyi established that "myosin" was really a complex of myosin and actin, tremendous progress has occurred in identifying and characterizing the contractile proteins and their interactions in muscle. These proteins are organized into nearly crystalline structures called myofibrils which occupy approximately 80% of the volume of the muscle cell. The myofibrils are responsible for force development and movement in higher organisms.

Muscle cells are more or less cylindrical in shape and contain on the order of a thousand myofibrils in cross-section. A phase contrast micrograph of a bovine skeletal muscle myofibril is shown in Fig. 1. The alternating dark (A bands) and light (I bands) are clearly visible. The I band is bisected by a phase-dense Z line. The repeating units (Z line to Z line) are termed sarcomeres. With myofibrils which have long sarcomeres, a less dense zone in the center of the A band can be seen which is called the H zone.

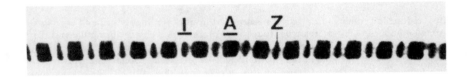

Figure 1. Phase contrast micrograph of a bovine skeletal muscle myofibril. Z-Z line; A-A band; I-I band. X 2500.

The origin of these banding patterns becomes apparent when the myofibril is observed in the electron microscope. Fig. 2 is a negatively stained, whole mount of a myofibril observed and photographed with a 1,000,000-volt electron microscope (conventional transmission electron microscopes have about one-tenth as much accelerating voltage and, therefore, much more difficulty in penetrating thick specimens). It can be observed that the A band region contains thick filaments while the I band has thinner diameter filaments. The two filament sets interdigitate and slide past each other as the sarcomere length changes during contraction and relaxation. The M line connects adjacent thick filaments in the middle of the sarcomere. Thin filaments are attached to the Z lines. An H zone, which does not contain thin filaments, can be observed in longer sarcomeres (Fig. 3). The thick and thin filaments

FOOD QUALITY • 183

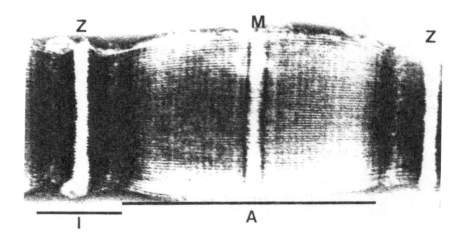

Figure 2. High-voltage electron micrograph of a negatively stained myofibril. A suspension of myofibrils was allowed to settle on a carbon coated grid, fixed with 2% glutaraldehyde in 100 mM KCl-20 mM potassium phosphate (pH 7.0) for 30 minutes, rinsed, and finally negatively stained with 2% uranyl acetate. X 54,000.

Figure 3. High-voltage electron micrograph of a myofibril which has been critical point-dried (see Ris, 1985 for procedures). Z-Z lines; M-M line; H-H zone. X 54,000.

are arranged in a hexagonal lattice (when viewed in cross-section) with each thin filament surrounded by three thick filaments and each thick filament encircled by six thin filaments. The visibility of thick filament profiles in the negatively stained preparations indicates the precision of packing since the microscope's depth of focus would encompass a large number of individual filaments. Fig. 4 demonstrates an exact superposition of thick filaments in the left A band and a slight twist of the lattice in the adjacent sarcomere to give a reduction in the apparent filament spacing.

Figure 4. High-voltage electron micrograph of a negatively stained myofibril. Myofibrils were processed as described in Fig. 2. The superimposed profiles of the thick filaments are more widely spaced in the left sarcomere (large arrow); the lattice is rotated in the one on the right (small arrow). The 43 nm repeat in the A band is visible (white arrows). X 54,000.

Myofibril Protein Composition and Protein-Protein Interactions

A list of the major myofibrillar proteins plus their estimated contents and sarcomere locations is shown in Table 1. Nine proteins constitute nearly 95% of the total. SDS polyacrylamide gels reveal no major unidentified bands (Fig. 5). Myosin and actin make up over half of the myofibrillar protein.

Table 1. Protein content and location in the myofibril

Component	Percent (w/w)	Location
Myosin	43	Thick filaments
Actin	22	Thin filaments
Titin (connectin)	8-10	Most of sarcomere?
Tropomyosin	5	Thin filaments
Troponin	5	Thin filaments
Nebulin	3-5	I band?
C-Protein	2	Thick filaments
M-Protein	2	M line
α-Actinin	2	Z line
β-Actinin	<1	End of thin filaments
γ-Actinin	<1	I band
Caldesmon	<1	I band
Enactin	<1	I band
Filamin	<1	Z line
Desmin	<1	Z line
Vinculin	<1	Z line
Talin	<1	Z line
Zeugmatin	<1	Z line
Z Protein	<1	Z line
Z-nin	<1	Z line
Myomesin	<1	M line
Creatine kinase	<1	M line
Skelemins	<1	M line
H Protein	<1	Thick filaments
X Protein	<1	Thick filaments
86 kd protein	<1	Thick filaments
I-Protein	<1	A-I junction

Table 2 summarizes the various protein complexes which form in the myofibril. The myosin-myosin and actin-actin interactions provide the backbone for the thick and thin filaments respectively. Approximately 300 myosin molecules are assembled in each thick filament, and they are packed such that a 43 nm repeat can be observed by x-ray diffraction or by electron microscopy (see Figs. 2, 4, and 5). About 400 actin monomers are polymerized into a twisted double strand as the core of the thin filament. A 38 nm repeat visible in electron micrographs arises from the troponin periodicity along the thin filaments (Fig. 6). Myosin heads bind to actin and go through a power

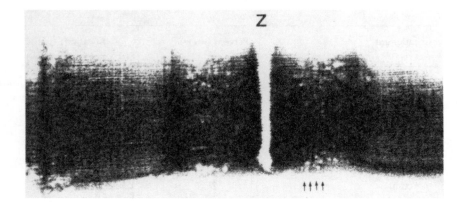

Figure 5. High-voltage electron micrograph of a negatively stained myofibril. Myofibrils were processed as described in Fig. 2. Note the periodic density profiles at 38 nm crossing the I band (arrows). X 54,000.

stroke during muscle contraction. They also bond tightly when ATP is depleted in muscle postmortem (rigor mortis). Tropomyosin and troponin respond to ionic calcium concentrations to regulate the myosin-actin interaction. Alpha-actinin and β-actinin bind to opposite ends of the actin filaments to attach them to the Z lines and to limit their length respectively. Creatine kinase, M-protein, and perhaps myomesin link the thick filaments together in a regularly spaced hexagonal lattice at the M line in the middle of the sarcomere.

Table 2. Myofibrillar protein-protein interactions

Proteins	Function
Myosin-Myosin	Thick filament assembly
Actin-Actin	Thin filament assembly
Myosin-Acti	Produce contractile force
Actin-Tropomyosin-Troponin	Regulation of contraction-relaxation
Actin-α-Actinin	Anchor thin filaments in Z line
Actin-β-Actinin	Cap the end of the thin filament
Myosin-Creatine Kinase-M-Protein-Myomesin	Anchor thick filaments together

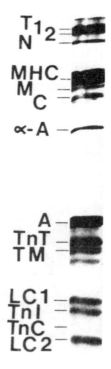

Figure 6. SDS polyacrylamide gel of bovine skeletal muscle myofibrils. Samples were run on a 10% separating gel at pH 9.3 using a procedure adapted from Giulian et al. (1983). Bands are labeled as follows: T1 and T2-titin; N-nebulin; MHC-myosin heavy chain; M-M protein; C-C protein; αA-α actinin; A-actin; TnT-troponin T; TM-tropomyosin; LC1-myosin light chain 1; TnI-troponin I; TnC-troponin C; LC2-myosin light chain 2.

Titin and Nebulin

The proteins titin and nebulin contribute more than 10% of the myofibrillar protein (Table 1) but as yet have not been fully characterized as to sarcomere location, function, or protein-protein interactions. Both proteins have extremely large subunit sizes as determined by SDS polyacrylamide gel electrophoresis. Titin, which is also called connectin, was discovered independently by Maruyama et al. (1976) and Wang et al. (1979). Its subunit size remains controversial with estimates ranging up to 2.8 million daltons (Maruyama et al., 1984; Kurzban and Wang, 1988). Nebulin is also very large with a subunit size of about 500,000 daltons (Wang and Williamson, 1980).

Antibody staining with immunofluorescence has been used to localize titin in the myofibril. Fig. 7 shows a bovine skeletal muscle myofibril stained with a polyclonal anti-titin. The major fluorescence is located at the A-I junction and at either side of the M line. Weak staining is also visible near the Z lines.

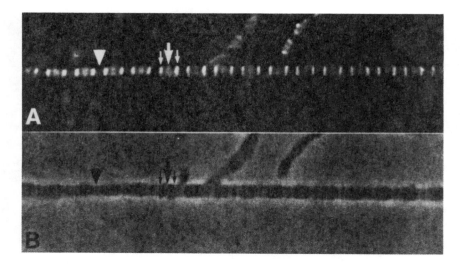

Figure 7. Immunofluorescence and phase micrographs of a bovine skeletal muscle myofibril stained with a polyclonal anti-titin antibody. Myofibrils were attached to glass slides, fixed in 0.1% glutaraldehyde, incubated with antibody, rinsed, incubated with fluorescein-labeled secondary antibody, and viewed after epifluorescent illumination (see Wang et al. 1988 for further details on the procedure). A. Immunofluorescence. Most intense staining is located near the A-I junctions (arrows). There is also staining near the middle of the A bands (wide arrow) and near the Z lines (arrowheads). B. Phase contrast. X 2500.

More definitive results have been obtained using monoclonal antibodies. Fig. 8 shows immunoblots of three different monoclonals developed in our laboratory which are specific for titin. Staining patterns using the antibodies are demonstrated in Fig. 9. The 9D10 antibody stains two zones per sarcomere near the A-I junction. The position of this staining changes with sarcomere length (Wang and Greaser, 1985). B5 binds most strongly at the Z lines. The E2 antibody gives most intense staining at the middle of the A band. High-voltage electron microscopy allows a more precise localization of antibody binding. Fig. 10 shows that the 9D10 contributes extra density in two bands per sarcomere in the I band region. These results suggest that there are two sets of titin molecules per sarcomere arranged symmetrically about the M line and extending to the Z line. A recent paper by Furst et al. (1988) using a more extensive group of monoclonals reached a similar conclusion.

In spite of the large quantity of titin in the myofibril, its structural localization by electron microscopy has proven difficult. Only in muscle stretched beyond thick and thin filament overlap are fine filaments visible which are now believed to contain titin (so called "gap filaments," see Sjostrand, 1962; Locker, 1984). The presence of some type of longitudinal connection is evident since even highly stretched myofibrils with no thick-thin filament overlap do not disintegrate (see Fig. 11). The titin filaments may function to prevent muscle damage by overstretching and help to stabilize the thick filaments in the center of the sarcomere (Horowits and Podolsky, 1987). They may also play a role in sarcomere assembly (Wang et al., 1988).

Much less is known about nebulin. The initial report indicated that nebulin was localized in the N2 line of the I band (Wang and Williamson, 1980). More recently, it has been suggested that the protein occupies a major portion of the I band (Wang and Wright, 1987). Nebulin is absent or present in very small amounts in cardiac muscle. Its function remains to be determined.

190 • BIOTECHNOLOGY

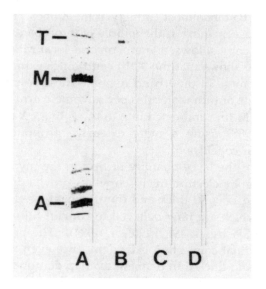

Figure 8. Immunoblots of titin monoclonals. Mice were immunized with purified bovine cardiac titin. Monoclonal antibodies were selected using ELISA tests and immunofluorescence staining patterns. The blotted myofibrillar protein was incubated with the appropriate primary antibody and processed using the procedures and reagents of Biorad Laboratories, Richmond, California. A. India ink stained nitrocellulose strip of blotted protein from an SDS gel. B. Monoclonal 9D10. C. Monoclonal 215B5. D. Monoclonal 225E2. Each monoclonal only stains the titin band. T-Titin; M-myosin heavy chain; A-actin.

Changes in Titin Postmortem

Attempts to identify postmortem proteolytic changes in the major myofibrillar proteins have been largely unsuccessful. No obvious changes in myosin, actin, tropomyosin, or α-actinin contents are readily apparent using SDS polyacrylamide gel electrophoresis. The troponin T subunit has been found to degrade to a 30,000-dalton fragment (Penny, 1976; Olson et al., 1977), but it is difficult to explain how such a change might weaken the myofibril structure and lead to increased tenderness with meat aging. Recent work suggests that titin and nebulin may be the most labile constituents of the myofibril to postmortem degradation (Lusby et al., 1983; Bandman and Zdanis, 1988). Work in our laboratory supports this conclusion. Fig. 12 shows an immunoblot of an SDS gel of bovine muscle samples obtained immediately after death (A) and at 48 hours postmortem (B-I). The blot was stained with a polyclonal anti-titin. It is clear that a significant proportion of the titin becomes fragmented into lower molecular weight pieces during the early postmortem period. There is also some variation in the extent of degradation in samples from different animals (lanes B-I). It is tempting to postulate that the degree of titin degradation may partially determine the fragility of the tissue and thus affect the tenderness of the cooked product.

Another approach to observing postmortem changes in titin is to examine antibody-stained myofibrils by immunofluorescence microscopy. The 9D10 antibody routinely stains two perpendicular zones per sarcomere in myofibrils near the A-I junction. Surprisingly, this pattern is altered in postmortem myofibrils from bovine psoas muscle (Ringkob et al., 1988). Fig. 13 shows the staining patterns of myofibrils obtained at 3 hours and 48 hours after death. In many myofibrils, the titin antibody stained 4 bands per sarcomere instead of two. Some myofibrils also showed staining throughout the I-band region or occurred in ragged patterns. These changes in staining are interpreted to result from the breakage of the long, elastic titin molecules due to proteolytic activity in postmortem muscle. Alternatively, the movement of part of the titin staining regions might occur from proteolysis or breakage of another protein to which titin is attached.

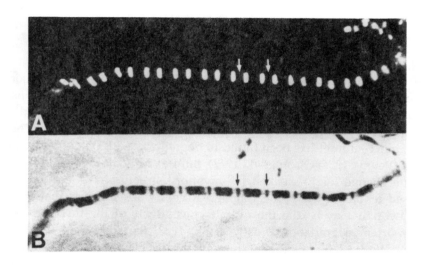
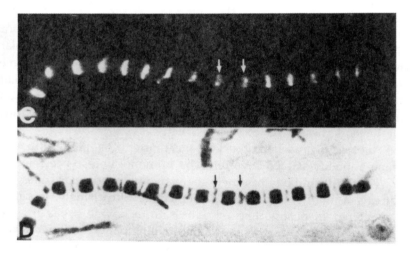

FOOD QUALITY • 193

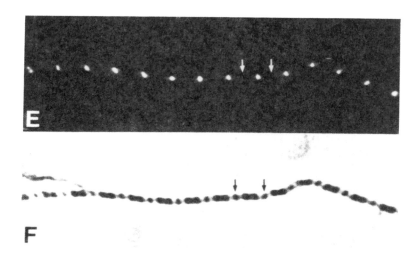

Figure 9. Immunofluorescence of myofibrils using monoclonal antibodies against titin. Myofibrils were processed using the different monoclonals as described in Fig. 7. Arrows indicate the positions of the Z lines. A. 9D10 staining. The antibody stains two perpendicular zones near the A-I junction. B. Phase contrast (same as A). C. B5 staining. The major staining occurs at the Z lines. D. Phase contrast (same as C). E. E2 staining. The antibody binds near the center of the A band. F. Phase contrast (same as E). X 2500.

194 • BIOTECHNOLOGY

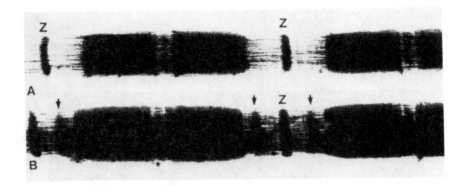

Figure 10. HVEM of myofibrils incubated with 9D10 anti-titin. Myofibrils (0.3 mg) were incubated with 0.03 mg of antibody for 1 hour at room temperature before being fixed and critical point-dried. A. Control myofibril. X 54,000. B. Antibody treated. Two perpendicular bands (arrows) of greater electron density are visible in the I band near the A-I junction. X 60,000.

Figure 11. Phase contrast micrograph of 48-hour postmortem bovine psoas myofibrils. Note that, although the thick and thin filament regions are stretched beyond overlap, the myofibril still maintains its longitudinal continuity. X 2500.

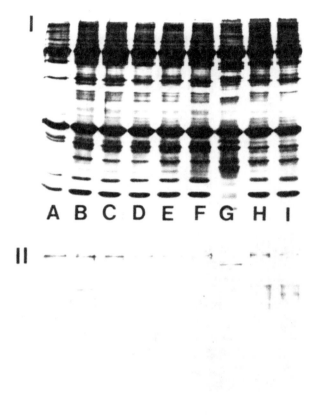

Figure 12. Immunoblots of 3-hour and 48-hour postmortem bovine longissimus myofibrils. I. India ink stained nitrocellulose sheet after blotting on SDS gel of 3-hour (lane A) and 48-hour (lanes B-I) samples. II. Polyclonal antibody staining of a companion blot with samples identical to those in I.

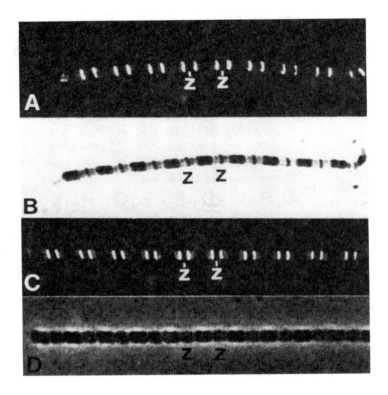

Figure 13. Immunofluorescence and phase contrast micrographs of bovine psoas myofibrils obtained from 3-hour and 48-hour postmortem muscle. A. Immunofluorescence. Myofibrils removed from the carcass at 1 hour after death were stained with 9D10 by procedures as outlined in Fig. 7. Two antibody stripes per sarcomere are visible. B. Phase contrast (same as A). C. Immunofluorescence, 48-hour postmortem. Four distinct antibody bands are visible on many of the myofibrils. D. Phase contrast (same as C). X 2500.

Salt concentrations such as those used in meat processing are known to alter the myofibril structure due to swelling and protein extraction (Offer and Trinick, 1983). They also alter the titin staining patterns of myofibrils (Fig. 14). The most common pattern is one in which antibody staining occurs in two zones on either side of the M line in addition to the A-I position. The salt apparently disrupts one or more of the titin anchor points to result in translocation of some of the titin epitopes.

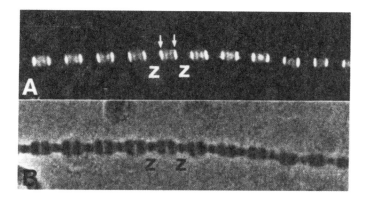

Figure 14. Immunofluorescence and phase contrast micrographs of bovine psoas myofibrils extracted with sodium chloride. A myofibril suspension was deposited on a glass coverslip and mixed with 0.5 M NaCl for 10 minutes. The excess solution was rinsed away with rigor buffer (Ringkob et al., 1988) and the myofibrils fixed and antibody stained using the 9D10 monoclonal as outlined in Fig. 7. A. Immunofluorescence. The antibody stains the A-I junctions (arrows) and two additional bands near the middle of the A band. B. Phase contrast. X 2500.

Current Knowledge and Future Prospects

We have attempted to provide an overview of the structure and protein constituents of muscle. An understanding of the molecular interactions of the contractile proteins should be the basis for technological improvements of muscle in its use for food. In spite of the vast information we have about muscle proteins and their function, the precise structural and/or proteolytic changes controlling meat tenderness remain to be determined. In addition, little is known about the effect of heating on protein-protein interactions as they affect meat quality. What effect does salt have on muscle proteins which are heat processed? Can we genetically engineer animals to contain larger amounts of key proteolytic enzymes to improve meat quality? Can altered bacterial strains be used to effect desired textural properties of fresh and processed meat? Both basic and applied research will be required to answer these key questions.

Acknowledgements

This work was supported by the College of Agricultural and Life Sciences, University of Wisconsin, Madison, Wisconsin, and by grants from the National Institutes of Health and the Beef Industry Council (Wisconsin Beef Council). The University of Wisconsin Hybridoma Facility's assistance in producing the monoclonal antibodies is also acknowledged.

References

Bandman, E. and Zdanis, D. 1988. An immunological method to assess protein degradation in post-mortem muscle. Meat Sci. 22: 1.

Furst, D. O., Osborn, M., Nave, R. and Weber, K. 1988. The organization of titin filaments in the half-sarcomere revealed by monoclonal antibodies in immunoelectron microscopy: a map of ten nonrepetitive epitopes starting at the Z line extends close to the M line. J. Cell Biol. 106: 1563.

Giulian, G. G., Moss, R. L. and Greaser, M. L. 1983. Improved methodology for analysis and quantitation of proteins on one-dimensional silver stained slab gels. Anal. Biochem. 129: 277.

Greaser, M. L. 1986. Conversion of muscle to meat. In "Muscle as Food," p. 37. Academic Press, New York.

Horowits, R. and Podolsky, R. J. 1987. The positional stability of thick filaments in activated skeletal muscle depends on sarcomere length: evidence for the role of titin filaments. J. Cell Biol. 105: 2217.

King, N. L. and Macfarlane, J. J. 1987. Muscle proteins. In Advances in Meat Research, Vol. 3, p. 21. Van Nostrand Reinhold Company, New York.

Kurzban, G. P. and Wang, K. 1988. Giant polypeptides of skeletal muscle titin: sedimentation equilibrium in guanidine hydrochloride. Biochem. Biophys. Res. Comm. 150: 1155.

Leavis, P. C. and Gergely, J. 1984. Thin filament proteins and thin filament-linked regulation of vertebrate muscle contraction. CRC Crit. Rev. Biochem 16: 235.

Locker, R. H. 1984. The role of gap filaments in muscle and in meat. Food Microstruct. 3: 17.

Lusby, M. L., Ridpath, J. F., Parrish, F. C., Jr. and Robson, R. M. 1983. Effect of postmortem storage on degradation of the myofibrillar protein titin in bovine longissimus muscle. J. Food Sci. 48: 1787.

Maruyama, K., Natori, R. and Nonomura, Y. 1976. New elastic protein from muscle. Nature 262: 58.

Maruyama, K., Kimura, S., Yoshidomi, H., Sawada, H. and Kikuchi, M. 1984. Molecular size and shape of β-connectin, an elastic protein of striated muscle. J. Biochem. 95: 1423.

Offer, G. and Trinick, J. 1983. On the mechanism of water holding in meat: the swelling and shrinking of myofibrils. Meat Sci. 8: 245.

Ohtsuki, I., Maruyama, K. and Ebashi, S. 1986. Regulatory and cytoskeletal proteins of vertebrate skeletal muscle. Adv. Prot. Chem. 38: 1.

Olson, D. G., Parrish, F. C., Jr., Dayton, W. R. and Goll, D. E. 1977. Effect of post-mortem storage and calcium activated factor on the myofibrillar proteins of bovine skeletal muscle. J. Food Sci. 42: 117.

Penny, I. F. 1976. The effect of conditioning on the myofibrillar proteins of pork muscle. J. Sci. Food Agric. 27: 1147.

Ringkob, T. P., Marsh, B. B. and Greaser, M. L. 1988. Change in titin position in postmortem bovine muscle. J. Food Sci. 53: 276.

Ris, H. 1985. The cytoplasmic filament system in critical point-dried whole mounts and plastic embedded sections. J. Cell Biol. 100: 1474.

Ruegg, J. C. 1988. "Calcium in Muscle Activation." Springer-Verlag, Berlin.

Sjostrand, F. 1962. The connections between A- and I-band filaments in striated frog muscle. J. Ultrastruct. Res. 7: 225.

Squire, J. M. 1986. Muscle: Design, Diversity, and Disease. Benjamin/Cummings Publishing Company, Menlo Park, California.

Wang, K., McClure, J. and Tu, A. 1979. Titin: major myofibrillar components of striated muscle. Proc. Nat. Acad. Sci. USA 76: 3698.

Wang, K. and Williamson, C. L. 1980. Identification of an N2 line protein of striated muscle. Proc. Nat. Acad. Sci. USA 77: 3254.

Wang, K. and Wright, J. 1987. Architecture of sarcomere matrix in skeletal muscle—evidence that nebulin constitutes a distinct set of nonextensible filaments in parallel with titin filaments. J. Cell Biol. 105: 27a.

Wang, S.-M. and Greaser, M. L. 1985. Immunocytochemical studies using a monoclonal antibody to bovine cardiac titin on intact and extracted myofibrils. J. Muscle Res. Cell Motil. 6: 293.

Wang, S.-M., Greaser, M. L., Schultz, E., Bulinski, J. C., Lin, J.-C. and Lessard, J. L. 1988. Studies on cardiac myofibrillogenesis with antibodies to titin, actin, tropomyosin, and myosin. J. Cell Biol. 107: 1075.

Molecular Components of Food: Repartitioning Agents or Hormones

Stephen B. Smith
Department of Animal Science
Texas A&M University
College Station, TX 77843, USA

The administration of repartitioning agents (the phenethanolamines and somatotropins) to livestock species causes such desirable effects as increases in protein concentration and a concomitant decrease in ether-extractable lipid in most muscles. These effects typically are associated with elevated muscle mass and a decrease in carcass adipose tissue mass. Some decrease in tenderness has been observed in meat from swine fed either class of repartitioning agent; this has not been of sufficient magnitude to be of practical concern. However, it is very clear that the phenethanolamines elicit significant reductions in tenderness in cattle, sheep and poultry. The decrease in tenderness, which persists in steers even after an extended withdrawal period, may be the result of alterations in myofiber diameter, decreases in the activities of endogenous proteases, or may in some manner be related to textural changes observed in meat from phenethanolamine-treated cattle and sheep.

Introduction

The feeding of grain to livestock and poultry species evolved in the U.S. for very practical reasons: grain typically is an inexpensive source of high quality feed; and the feeding of grain ("finishing") increases growth rate. In cattle, grain feeding also increases the palatability and acceptability of beef (Zinn et al., 1970; Tatum et al., 1980). Longissimus muscle from cattle fed grain contains greater amounts of marbling, and the concomitant increase in muscle lipid may be a major contributor to the increased palatability of grain-fed beef (Smith et al., 1983).

The system for grading beef in the U.S. recognizes the relationship between grain feeding, marbling and animal maturity. To achieve the level of marbling required to grade USDA Choice at a youthful maturity, cattle must be fed diets high in grain. Currently, the difference in value between USDA Choice beef carcasses and those grading one level lower (USDA Select) is approximately $50.00/carcass. A major problem in the grain feeding of beef cattle is that it takes about 30% carcass fat before the deposition of 5% fat in the longissimus muscle (typical of USDA Choice meat) is realized (Tatum et al., 1980). The result is that excess fat must be trimmed off of the retail cuts prior to their sale to the consumer. Although the grain feeding of other meat-bearing species may have evolved for reasons other than achieving a specific degree of marbling, the result is the same: too much carcass fat by the time the animals reach typical slaughter weight.

To combat the excessive deposition of carcass fat without reducing (or perhaps enhancing) feed efficiency and/or growth rate, a family of compounds has been introduced which, when fed to or injected in livestock, increases muscle growth and reduces carcass fat accretion. Hence, these compounds are collectively referred to as repartitioning agents. Two classes of repartitioning agents have been characterized recently: the somatotropins (growth hormones), both pituitary-derived and their recombinant analogs; and the synthetic phenethanolamines (which include the β-adrenergic agonists).

The injection of pigs with somatotropin increases muscle growth and decreases adipose tissue accretion (Machlin, 1972; Chung et al., 1985; Etherton et al., 1986, 1987; Evock et al., 1988). Similarly, the feeding of phenethanolamines such as clenbuterol, cimaterol or ractopamine to livestock species typically results in marked muscle hypertrophy and a reduction in carcass fat. This has been demonstrated clearly for cattle (Ricks et al., 1984; Miller et al., 1987), sheep (Baker et al., 1984; Hamby et al., 1986; Kim et al., 1987) and swine (Jones et al., 1985; Moser et al., 1986).

The increased production of lean meat through the use of repartitioning agents certainly is in keeping with current consumer demands and industry trends. However, the production of lean meat through the use of these compounds may be eliciting side effects that will reduce both their acceptance and the potential profits that would accrue to the producer through their use. If these compounds reduce levels of marbling in beef, then carcasses from the treated animals will be lower in value. Alternatively, repartitioning agents may reduce meat quality, thereby decreasing consumer acceptability.

The marked effects of repartitioning agents on muscle and adipose tissue growth indicate that a variety of cellular processes are altered dramatically when these agents are administered to livestock.

Along with the beneficial results of greater muscle protein accretion and reduced adipose tissue, other less desirable cellular/metabolic processes may be elicited that have negative effects on carcass and meat quality. The possible modes of action of repartitioning agents on cellular growth must be elucidated to provide a better understanding of how these compounds could potentially affect the quality of meat.

Mechanism(s) of Action

Several recent reports have aided in the elucidation of the potential mechanism(s) of action of repartitioning agents in the target species, sheep, cattle swine and poultry. Most of these reports deal with the cellular actions of the phenethanolamines, and substantially less information is available for the somatotropins. However, the phenethanolamines may be working indirectly via elevated somatotropin secretion. Cimaterol elevates circulating somatotropin in growing lambs (Beermann et al., 1987), and clenbuterol stimulates the secretion of somatotropin in cultured bovine adenohypophyseal cells (Welsh et al., 1987). Whether the phenethanolamines modify carcass composition by increasing circulating somatotropin levels, or actually interact with the target tissues directly, it is possible that the phenethanolamines and the somatotropins have common mechamisms of action in their effects on the quality of meat.

The increase in muscle mass elicited by repartitioning agents clearly is the result of myofiber hypertrophy. However, there is some question as to whether the phenethanolamines affect type I (oxidative) myofibers, type II (glycolytic) myofibers, or both. Hamby et al, (1986) reported that longissimus muscle strips obtained from clenbuterol-fed lambs exhibited greater rates of glycogen synthesis and overall glucose utilization *in vitro* than muscle from untreated lambs, suggesting larger type II myofibers in the treated lambs. This subsequently was corroborated by Kim et al., (1987), who observed hypertrophy only for the type II fibers of cimaterol-fed lambs. In contrast, Beermann et al., (1987) reported 30% increases in both type I and type II myofibers from semitendinosus muscle of lambs fed cimaterol. In mature heifers, clenbuterol treatment (10 mg·head^{-1}·d^{-1}) resulted in hypertrophy to type II longissimus muscle fibers only (Miller et al., 1987), while in young steers, clenbuterol given at a lesser dose (7 mg·head^{-1}·d^{-1}) elicited hypertrophy of both type I and type II longissimus muscle myofibers

(Garcia et al., 1988; Fig. 1). The sum of these investigations is that the administration of phenethanolamines consistently increases the rate of hypertrophy of the glycolytic type II fibers, and in some instances increases the hypertrophy of the aerobic type I fibers.

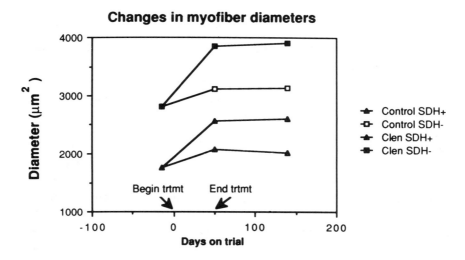

Figure 1. Changes in succinate dehydrogenase (SDH) positive (oxidative) and SDH negative (glycolytic) myofiber diameters with clenbuterol treatment. Clenbuterol (Clen) was fed to Angus steers for 50 d (beginning on d 0); on d 51, clenbuterol was withdrawn from the feed of the treated animals. Eight steers from each treatment group were slaughtered at each of the indicated time points (from Garcia et al., 1987).

The hypertrophy of muscle involves a balance between the processes of protein synthesis and protein degradation. Protein synthesis in muscle involves transcription of the genes encoding myofibrillar proteins as well as translation of the RNA transcripts. Degradation of protein is assumed to involve the proteolytic activities of both the cathepsins and the calcium-dependent proteases. Feeding trials with sheep demonstrated that clenbuterol did not increase the fractional rate of protein synthesis (Bohorov et al., 1987), suggesting that this agent increases muscle mass by decreasing the rate of myofibrillar protein degradation. Correspondingly, Wang and Beermann (1988) reported that dietary cimaterol reduced the activity of the calcium-dependent proteases in ovine muscle.

Indirect evidence exists to indicate that the phenethanolamines also increase muscle mass by stimulating protein synthesis. Beermann et al., (1987) reported that cimaterol increased the concentration of RNA in the muscle of growing lambs. Similarly, clenbuterol treatment resulted in elevated levels of the mRNA encoding actin and the fast isoform of myosin light chain-1 (Garcia et al., 1988; Fig. 2) in the longissimus muscle of steers. These results suggest that phenethanol-amines work, in part, by increasing the rate of myofibrillar protein synthesis.

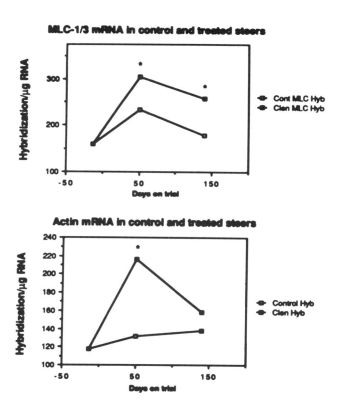

Figure 2. Changes in the concentration (hybridization units/µg RNA) of the mRNAs encoding myosin light chain-1/3 (MLC-1/3) and actin. Levels of hybridization were determined in a commercial slot-blot apparatus. Details of the experimental design are given in Fig. 1 (from Garcia et al., 1988 and unpublished observations from the author's laboratory).

Effects on Meat Quality

The marked reduction the lipid content of meat elicited by repartitioning agents potentially could reduce the overall quality of meat obtained from treated animals. Somatotropins reduce the lipid content of pork to as little as 1% (Beermann et al., 1988), while the phenethanolamines commonly reduce pork, lamb and beef muscle lipid by 10 to 25% (Baker et al., 1984; Ricks et al., 1984; Jones et al., 1985; Hamby et al., 1986). However, juiciness of cooked pork chops was not affected by somatropin treatment (Evock et al., 1988). Because juiciness is the attribute most affected by lipid content, it is unlikely the reducing the lipid content of pork through the use of repartitioning agents will have a major effect on meat quality. The reduction of lipid in beef elicited by phenethanolamines may have a more pronounced effect on the palatability of beef because, as mentioned previously, intramuscular fat content of bovine longissimus muscle is highly correlated with beef flavor desirability rating (Smith et al., 1983).

A more serious problem associated with the treatment of animals with repartitioning agents is the reduction in tenderness elicited by these compounds. Hamby et al. (1986) and Lee et al. (1988) reported that β-adrenergic agonists (clenbuterol and cimaterol, respectively) substantially reduced the tenderness of meat from growing lambs. Similarly, Miller et al. (1987) reported increased Warner-Bratzler shear force values for longissimus muscle from clenbuterol-treated heifers, which also was observed in muscle from treated steers (Table 1). The data in Table 1 also indicate increases in meat coarseness with clenbuterol treatment.

While the phenethanolamines appear to have little effect on pork quality, pituitary-derived procine somatotropin caused a reduction in the taste panel-evaluated tenderness of pork (Fig. 3, derived from data of Evock et al., 1988). This reduction in pork tenderness may be of little practical concern, because only the recombinant form of somatotropin will be used in production.

Although repartitioning agents may affect both protein synthesis and degradation, it is likely that the latter process has the greater influence over meat tenderness. As indicated above, clenbuterol decreases the tenderness of lamb (Hamby et al., 1986), and Wang and Beermann (1988) reported a reduction in calcium-dependent protease activity in muscle from cimaterol-treated lambs. The tenderness of chicken breast muscle also was reduced by cimaterol treatment (Morgan et al., 1988a,b), and these investigators reported reduced cathepsin B activity in the breast muscle of treated chickens.

Table 1. Carcass composition and quality characteristics of steers fed clenbuterol for 50 d and after 90d withdrawal

| Days on Trial | 0 | 50 | | 140 | |
Observation	Initial	Control	Clen	Control	Clen
Actual fat thickness, cm	.36^a	.78^b	.74^b	1.1^c	1.1^c
Adjusted fat thickness, cm	.36^a	.86^b	.78^b	1.2^c	1.2^c
Marbling Score	PD^{86a}	T^{98b}	T^{32a}	MT^{15d}	Sl^{89c}
Quality grade	St^{0a}	St^{+b}	St^{0a}	Ch^{0d}	Se^{+c}
Shear force, kg	5.7^c	5.9^c	7.0^d	2.9^a	3.7^b
Lean firmness score^e	5.9^a	6.5^a	7.6^b	6.1^a	7.0^{ab}
Lean texture score^f	4.9^{ab}	6.3^b	4.9^{ab}	6.5^b	3.4^a

[a,b,c,d]Means in the same row whose superscripts differ are different (P<.05).
[e]Scored: 7=very firm; 4=moderately firm; 1=extremely soft.
[f]Scored: 7=extremely fine; 4=slightly coarse; 1=extremely coarse.

It is a typical procedure in the U.S. to age carcasses at 4°C for 7 to 10 days postmortem, a process that results in increased meat tenderness. Either the cathepsins or the calcium-dependent proteases, or both, could be involved in the postmortem tenderization of meat (Dutson and Lawrie, 1974; Goll et al., 1983; Koohmaraie et al., 1988), primarily through their effects on specific structures within the myofiber (for a complete review, see Goll et al., 1983). Thus, a reduction in the activities of the proteases endogenous to muscle could eliminate or reduce the tenderizing effects of the aging of meat.

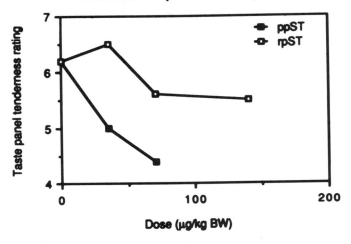

Figure 3. Taste panel tenderness of pork loin chops from pigs treated with increasing dosages of pituitary-derived porcine somatropin (ppST) or recombinant porcine somatotropin (rpST). 0=dislike extremely; 10=like extremely (From Evock et al., 1988).

Investigations of the role of proteases in the postmortem tenderization of meat typically have concentrated on their effects on myofibrillar proteins. However, these proteases also are involved in the turnover of connective tissues, and lower protease activities as a result of treatment with repartitioning agents would allow newly formed collagens to mature to form the less soluble, highly crosslinked connective tissues (Bailey, 1988). This would be aggravated if protein synthesis (to include collagen synthesis) is elevated in animals treated with repartitioning agents, as the data suggest. The coarseness of meat from clenbuterol-treated steers, which persists (along with elevated toughness) even after 90 days withdrawal of the compound from the feed (Table 1), could be the result of elevated levels of more mature, extensively crosslinked collagen in meat from the treated animals. Hence, one of the factors associated with enhanced muscle growth (i.e., decreased protein degradation) could be causative in reducing the tenderness of meat from sheep and cattle treated with repartitioning agents.

Fiber type composition also may influence tenderness of meat from animals treated with repartitioning agents. Calkins et al. (1981) reported that a greater percentage of oxidative fibers in beef was correlated with greater tenderness, although they did not provide a cellular basis for this observation. Because repartitioning agents have been demonstrated in some cases to increase type II fibers selectively (Miller et al., 1987; Lee et al., 1988), the greater proportion of glycolytic fibers may result in less tender meat. However, the demonstration of decreased tenderness in beef in which all myofiber types increased proportionately (Garcia et al., 1988) would appear to rule out fiber type composition as the basis for the alterations in meat tenderness caused by phenethanolamines.

Effects on Carcass Quality

The increased longissimus muscle (ribeye) cross-sectional area and reduced carcass fatness of cattle and sheep resulting from treatment with repartitioning agents typically results in improved USDA yield grades (Baker et al., 1984; Ricks et al., 1984; Miller et al., 1987). With the reduction in overall carcass fat, reductions in intramuscular (marbling) fat would be anticipated in cattle. Ricks et al. (1984) observed small reductions in marbling scores in feedlot cattle fed clenbuterol, but the effect was not statistically significant. In contrast, Miller et al. (1987) reported reductions in marbling scores in cattle fed clenbuterol sufficient to reduce USDA quality grades by one full grade in heifers; we have observed virtually identical results in steers (Table 1). The finding in heifers was accompanied by concomitant decreases in intramuscular adipocyte volumes (Miller et al., 1987). In steers, the effect on marbling scores persisted even after a 90-day withdrawal from treatment (Table 1).

Because of the importance of marbling scores in the determination of carcass quality grade for cattle, the use of the phenethanolamine, clenbuterol, would result in a substanial reduction in carcass value. Similar carcass data for the other phenethanolamines or for the somatotropins are not yet available, so it is uncertain if this effect will be limited to clenbuterol, or will be a generalized effect of repartitioning agents. In lambs and pigs, in which marbling scores are not determinants of carcass quality, repartitioning agents will not result in lowered carcass quality.

Conclusions

In summary, repartitioning agents, to include the somatotropins and phenethanolamines, increase lean growth in swine, but have no appreciable effect on pork quality. In contrast, beef carcass USDA quality grade is reduced significantly by the phenethanolamines, primarily through a reduction in marbling scores. Tenderness of beef and lamb also is affected negatively, which may be the result of decreased endogenous protease activity. Thus, it remains to be demonstrated that the increase in lean yield in animals treated with repartitioning agents will offset the potential reduction in carcass and/or meat quality, particularly for ruminant species.

References

Bailey, A.J. 1988. Connective tissue and meat quality. Proc. Internat. Cong. Meat Sci. Tech. 34: 152.

Baker, P.K., Dalrymple, R.H., Ingle, D.L. and Ricks, C.A. 1984. Use of a β-adrenergic agonist to alter muscle and fat deposition in lambs. J. Anim. Sci. 59: 1256.

Beermann, D.H., Butler, W.R., Hogue, D.E., Fishell, V.K., Dalrymple, R.H., Ricks, C.A. and Scanes, C.G. 1987. Cimaterol-induced muscle hypertrophy and altered endocrine status in lambs. J. Anim. Sci. 65: 1514.

Beermann, D.H., Armbruster, G., Boyd, R.D., Roneker, K. and Fagin, K.D. 1988. Comparison of the effects of two recombinant forms of porcine somatotropin (pST) on pork composition and palatability. J. Anim. Sci. 66 (Suppl. 1): 281. (Abstract).

Bohorov, O., Buttery, P.J., Correia, J.H.R.D. and Soar, J.B. 1987. The effect of the β-2-adrenergic agonist clenbuterol or implanatation with oestradiol plus trenbolone acetate on protein metabolism in wether lambs. Br. J. Nutr. 57: 99.

Calkins, C.R., Dutson, T.R., Smith, G.C., Carpenter, Z.L. and Davis, G.W. 1981. Relationship of fiber type composition to marbling and tenderness of bovine muscle. J. Food Sci. 46: 708.

Chung, C.S., Etherton, T.D. and Wiggins, J.P. 1985. Stimulation of swine growth by procine growth hormone. J. Anim. Sci. 60: 118.

Dutson, T.R. and Lawrie, R.A. 1974. Release of lysosomal enzymes during post mortem conditioning and their relationship to tenderness. J. Food Technol. 9: 43.

Etherton, T.D., Wiggins, J.P., Chung, S.C., Evock, C.M., Rebhun, J.F. and Walton, P.E. 1986. Stimulation of pig growth performance by porcine growth hormone and growth hormone-releasing factor. J. Anim. Sci. 63: 1389.

Etherton, T.D., Wiggins, J.P., Evock, C.M., Chung, C.S., Rebhun, J.F., Walton and Steele, N.C. 1987. Stimulation of pig growth performance by porcine growth hormone: Determination of the dose-response relationship. J. Anim. Sci. 64: 433.

Evock, C.M., Etherton, T.D., Chung, C.S. and Ivy, R.E. 1988. Pituitary porcine growth hormone (pGH) and a recombinant pGH analog stimulate pig growth performance in a similar manner. J. Anim. Sci. 66: 1928.

Garcia, D.K., Davis, S.K. and Smith, S.B. 1988. Specific gene expression in longissimus muscle of Angus steers fed clenbuterol. Proc. Recip. Meat Conf. 40: 155.

Goll, D.E., Otsuka, Y., Nagainis, P.A., Shannon, J.D., Sathe, S.K. and Muguruma, M. 1983. Role of muscle proteinsases in maintenance of muscle integrity and mass. J. Food Biochem. 7: 137.

Hamby, P.L., Stouffer, J.R. and Smith, S.B. 1986. Muscle metabolism and real-time ultrasound measurement of muscle and subcutaneous adipose tissue growth in lambs fed diets containing a beta-agonist. J. Anim. Sci. 63: 1410.

Jones, R.W., Easter, R.A., McKeith, F.K., Dalrymple, R.H., Maddock, H.M. and Bechtel, P.J. 1985. Effect of the ß-adrenergic agonist cimaterol (CL 263,780) on the growth and carcass characteristics of finishing swine. J. Anim. Sci. 61: 905.

Kim, Y.S., Lee, Y.B. and Dalrymple, R.H. 1987. Effect of the repartitioning agent cimaterol on growth, carcass and skeletal muscle characteristics in lambs. J. Anim. Sci. 65: 1392.

Koohmaraie, M., Seideman, S.C., Schollmeyer, J.E., Dutson, T.R. and Babiker, A.S. 1988. Factors associated with the tenderness of three bovine muscles. J. Food Sci. 53: 407.

Lee, Y.B., Jung, H., Kim, Y.S. and Dalrymple, R.H. 1988. Effect of cimaterol (CL 263,780) on meat quality in lambs. J. Anim. Sci. 66 (Suppl. 1): 279. (Abstract).

Machlin, L.J. 1972. Effect of porcine growth hormone on growth and carcass composition of the pig. J. Anim. Sci. 35:794.

Miller, M.F., Garcia, D.K., Coleman, M.E., Ekeren, P.A., Lunt, D.K., Wagner, K.A., Procknor, M., Welsh, T.H., Jr. and Smith, S.B. 1987. Adipose tissue, longissimus muscle and anterior pituitary growth and function in clenbuterol-fed heifers. J. Anim. Sci. 66: 12.

Morgan, J.B., Calkins, C.R. and Jones, S.J. 1988a. Cimaterol-fed broiler chickens: Changes in tenderness, cathepsin B activity and composition. J. Anim. Sci. 66 (Suppl. 1): 278. (Abstract).

Morgan, J.B., Jones, S.J. and Calkins, C.R. 1988b. Cimaterol-fed broiler chickens: Influence on muscle protein turnover. J. Anim. Sci. 66 (Suppl. 1): 278. (Abstract).

Moser, R.L., Dalrymple, R.H., Cornelius, S.G., Pettigrew, J.E. and Allen, C.E. 1986. Effect of cimaterol (CL 263,780) as a repartitioning agent in the diet for finishing pigs. J. Anim. Sci. 62: 21.

Ricks, C.A., Dalrymple, R.H., Baker, P.K. and Ingle, D.L. 1984. Use of a ß-agonist to alter fat and muscle deposition in steers. J. Anim. Sci. 59: 1247.

Smith, G.C., Savell, J.W., Cross, H.R. and Carpenter, Z.L. 1983. The relationship of USDA quality grade to beef flavor. Food Techn. May: 233.

Tatum, J.D., Smith, G.C., Berry, B.W., Murphey, C.E., Williams, F.L. and Carpenter, Z.L. 1980. Carcass characteristics, time on feed and cooked beef palatability attributes. J. Anim. Sci. 50: 833.

Wang, S.-Y. and Beermann, D.H. 1988. Reduced calcium-dependent proteinase activity in cimaterol-induced muscle hypertrophy in lambs. J. Anim. Sci. 66: 2545.

Welsh, T.H., Jr., Smith, S.B., Sutton, M.R. and Wagner, K.A. 1987. Growth hormone releasing factor and clenbuterol regulation of bovine growth hormone secretion *in vitro*. J. Animl. Sci. 65 (Suppl. 1):279. (Abstract).

Zinn, D.W., Gaskins, C.T., Gann, G.L. and Hedrick, H.B. 1970. Beef muscle tenderness as influenced by days on feed, sex, maturity and anatomical location. J. Anim. Sci. 31: 307.

Summary

There is a pattern to the way knowledge of biological systems has accumulated. Physical and microscopical observations give an initial appraisal and definition of a problem. The matter is then defined in chemical terms. The chemical description inspires a biochemical study that aims to elucidate which enzymes and which regulatory factors are pertinent. In favorable circumstances, recombinant DNA technology may be used to modify the usual biochemical changes or to test the theories of the enzymologists. This pattern of biological research is nicely illustrated in the contributions to this section.

The unifying theme is the texture of foods using the examples of fruit firmness in tomatoes and toughness in meat. Seeking to understand the components of a desirable texture, Dr. Gross has explored the chemistry of the tomato fruit cell wall, and Dr. Greaser the intricate molecular detail of myofibrillar structure of muscle. Each chemical study leads to a postulate of the enzymic events that contribute textural differences. The enzymes are not considered here in detail, but their activities and how they are regulated are in focus. Dr. Bleecker's paper on ethylene as a regulator of plant development, and Dr. Smith's paper on the effects of the administration of somatotropin and phenethanolamines on muscle development are major contributions in this context.

Dr. Gross's paper, too, leads us directly to a consideration of ethylene physiology when he shows that a sub-fraction of the cell wall, present only as ripening is initiated, has the capacity to elicit ethylene production when added to non-ripening fruit. There are now many reports of cell wall derived elicitors of ethylene synthesis but this one demands attention for it makes a timely appearance just before the tissue normally produces large amounts of ethylene. Apart from its enormous practical importance in agriculture, and especially in horticulture, there is a fascination in the very notion of ethylene, a small volatile molecule, as a regulator of plant growth and development. The pathway of ethylene biosynthesis has been known for some time, largely from the work of Professor Shang Fa Yang. However, the isolation of the rate limiting biosynthetic enzymes has proven very difficult. Likewise, clear definition that ethylene binds to a specific receptor or receptors has been difficult to achieve. Dr. Bleecker has filled both these voids. With the help of monoclonal antibodies, he has purified and characterized an ACC snythase protein from wounded tomato fruit tissue. This opens the way to sequence data and, in time, gene definition and beyond that gene regulation. Dr. Bleecker also shows the way to an *Arabidopsis* gene that may code for an ethylene receptor. The contributions raise exciting prospects, and the commercial possibilities that arise from Dr. Bleecker's work are large and broad indeed.

The commercial prospects include the production of transgenic plants with altered control of ethylene production or ethylene perception. The usefulness of modifying plants by single gene changes is shown elegantly by Dr. Bennett, who described work done jointly by groups at the University of California at Davis and Berkely. The introduction in tomato of the structural gene for the pectin-hydrolyzing enzyme, polygalacturonase, regulated by a proplyene-sensitive promoter that was active in the "ripening inhibited" mutant allowed an assessment of the role that endopolygalacturonase plays in ripening. The transgenic plants accumulated the pectin-splitting enzyme in the expected tissues and at the expected time. The appearance of the enzyme resulted in an increase in the proportion of the cell wall that was extracted by a strong chelator. However, the phenotype of the transgenic line was unaltered in terms of ethylene production, lycopene accumulation or the softening of the fruit. This sequence of experiments appears to finally disprove the hypothesis of Tigchelaar et al. (1978) that endopolygalacturonase activity has a pivotal role in the regulation of ripening. Since the enzyme mediated change in the properties of the uronic acid polymers in the cell wall did not result in a measurable change in texture in the fruit of the transgenic plants, alternative or additional components that regulate softening will be sought. This will not surprise plant breeders who have recognized that softening of tomatoes is pleitropically regulated.

Some of the difficulty of interpreting experiments on texture stem from the limited information available on the chemistry of the structural components. Dr. Gross refers to this as concerns the cell walls of fruit, and Dr. Greaser highlights it in his consideration of myofibrillar structure. To probe the complexity of these structures a battery of methods are needed. Dr. Greaser has explained how these methods can probe the molecular interactions of huge and complex molecules. His use of immunology has led to quite new insights into myofibril structure and function and allowed the formulation of hypotheses of how post mortem changes lead to a deteriation in meat quality. In time, immunology may also help to elucidate plant cell wall structure.

Recombinant DNA technology has increased the availability at reasonable cost of a number of physiologically active chemicals. Dr. Smith's paper describes how the use of two classes of chemicals in livestock feeding results in an increase in protein and a decrease in fat in muscle tissues. This desirable change appears to be associated with a decrease in tenderness. Dr. Smith postulates that a decrease in protein turnover, reflecting a decrease in endogenous proteases, contributes to the loss of tenderness. So, we are brought back at the end of this session, to a point that Dr. Gross emphasised at its beginning. In trying to understand changes in the physical properties of supramolecules, such as muscle myofibrils or cell walls, the interplay of synthetic and degradative forces need to be considered; the interplay is such that a change in one component seldom leaves the other unchanged.

Reference

Tigchelaar, E.C., McGlasson, W.B. and Buescher, R.W. 1978. Genetic regulation of tomato fruit ripening. Hort. Science 13:508.

C.J. Brady
CSIRO
Division of Food Research
North Ryde, Australia

EVALUATION OF
FOOD QUALITY

Model Non-Isotopic Hybridization Systems for Detection of Foodborne Bacteria: Preliminary Results and Future Prospects

S.W. Chan
S. Wilson
H-Y Hsu
W. King
D.H. Halbert
J.D. Klinger
GENE-TRAK Systems
Framingham, MA 01701, USA

New information on the chemistry of nucleic acids, nucleic acid hybridization, and genetic relatedness of microorganisms based on DNA and RNA have provided the means to develop rapid, specific, and sensitive assays for product contamination. Such assays can have relatively broad specificity (e.g., all members of a given species), or be highly restricted (e.g., detecting the presence of a virulence-related gene) depending on the choice of nucleic acid target, test conditions and product requirements. Increasing information on chromosomal and ribosomal sequences from a wide variety of organisms has allowed a more systematic approach to probe selection and development. These principles are illustrated in the recent development of rapid colorimetric hybridization assays for *Salmonella*, *E. coli*, and *Listeria* in food. Recent progress in detection and separation procedures has allowed use of non-isotopic methods, and rapid (2 day) test formats. Preliminary results suggest that for most samples, these assays have excellent inclusivity and exclusivity, and sensitivity and specificity at least equivalent to routine culture methods. Levels of product contaminants are often low, requiring cultural enrichment to reach detection limits of current hybridization assays. However, in the future, amplification of target nucleic acids and signal in hybridization assays will provide sensitivity for detection of extremely low numbers of organisms (e.g., 1-100). Data is presented on the performance of new simplified non-isotopic tests emphasized detection of *Salmonella*.

Introduction

In the last five years we have witnessed increased interest in microbial contamination of food products and the potential for foodborne outbreaks. Such concern has been justified as *Salmonella, Campylobacter, Yersinia, Listeria,* and pathogenic *E. coli* have been implicated in both sporadic and widespread instances of foodborne illness during this period. These occurrences have brought into question many standard approaches to the manufacture and testing of commercial food products. Public awareness has also been heightened by media attention to outbreaks of *Salmonellosis* and *Listeriosis*.

Wide-ranging approaches have been proposed for reduction of foodborne microbial contamination at levels of production, distribution, preparation and storage of at-risk products. Examples include development of poultry production methods which would reduce *Salmonella* carriage, strict adherence to policies of separation of raw from finished materials in dairy processing environments, evaluation of more effective methods and programs for cleaning and disinfection, and introduction of processing machinery which can be operated in an aseptic manner. One of the most far-reaching approaches involves analysis of critical control points in manufacturing procedures with monitoring focused on these important steps. Implicit in any of these approaches is the need for rapid, definitive testing methods for the pathogens in question. Traditional methods have tended to be tedious and time-consuming. Additionally, in the absence of sensitive comparative methods it has been difficult for users to quantitate actual rates of false positive and false negative results and the costs and impact associated with these incorrect reports.

All of these factors have led to a brisk level of academic and commercial interest in developing new approaches more rapid and accurate to testing for foodborne pathogens. Examples include hydrophobic grid membrane techniques, impedance microbiology, ELISA, labeled bacteriophage methods, HPLC analysis of cell wall components, and nucleic acid hybridization. Each of these methods has inherent strengths and weaknesses, or technical limitations for practical implementation.

We have concentrated exclusively on development and refinement of nucleic acid hybridization methods for detection of pathogenic organisms in food and clinical specimens. Several attributes of hybridization assays make them extremely appealing for diagnostic procedures. Nucleotide sequences will form double strands by hydrogen bonding only to the specific complimentary sequence. The specificity of these interactions is influenced by sequence homology and reaction conditions such as temperature, ionic strength, and pH. By labeling this sequence, the extent of the reaction can be monitored, and the presence of the complementary target strand in an unknown sample is thus determined. Such labeled probe sequences are generally DNA and can be tagged by incorporating isotopically or chemically modified bases; frequent examples of labels include ^{32}P, ^{125}I, ^{35}S, fluorescein, biotin, chemiluminescent moieties, etc. Target molecules include organism-specific genomic DNA, ribosomal RNA (rRNA), messenger RNA (mRNA), or plasmid DNA. Presence of genomic targets (on a per cell) thereby affording additional assay sensitivity. It is not necessary to know the function of target (or probe) sequences, but rather to simply verify that they are found exclusively in the organism of interest.

Hybridization techniques have been used in research settings for many years but the procedures were overly complicated for application by non-specialists. The hybridization assay developed by GENE-TRAK Systems to detect *Salmonella* in foods was one of the earliest assay systems designed to overcome these difficulties. The assay uses rather large (approximately 1 Kb) ^{32}P-labeled DNA probes homologous to chromosomal regions found in all *Salmonella* serovars tested (Fitts, 1985). In 1987, GENE-TRAK also introduced a hybridization assay in a simple format to detect *Listeria* species in foods using ^{32}P-labeled oligonucleotide probes to detect unique regions of *Listeria* 16S rRNA (Klinger et al., 1988). The *Salmonella* assay has undergone AOAC collaborative evaluation (Flowers et al., 1987), and has been shown to be more sensitive than reference microbiological methods (Flowers et al., 1987). The GENE-TRAK *Salmonella* assay, and assays for heat-stable and heat-labile genes of *E. coli* are the three hybridization procedures for which collaborative studies have been completed and are included as official methods in the AOAC Bacteriological Analytical Manual (AOAC, 1984; AOAC, 1987).

The hybridization assays mentioned thus far are superior to conventional methods in terms of speed, ease of use, and sensitivity. However, in order to achieve adequate levels of sensitivity, isotopic labeling has been required. The use of isotopes is acceptable in many testing environments, but not desirable in others (e.g., certain instances in which a food testing laboratory is physically located within a food manufacturing facility). Isotopic labels have therefore been at least a partial barrier to widespread introduction of hybridization methods.

Very recently, simple, sensitive, non-isotopic colorimetric methods for detection of several foodborne organisms have been developed in our laboratories. In this paper we will discuss the development of these assays, as model systems, with emphasis on detection of *Salmonella* and selective examples on *E. coli.*

Principles of the colorimetric DNA hybridization assay
The GENE-TRAK colorimetric hybridization assays employ organism-specific DNA probes labeled with a non-radioactive compound, to detect foodborne pathogens. In general, food samples are enriched in appropriate broth media, and the nucleic acids are released by any of a number of simple chemical lysis agents. Hybridization is carried out with two different probes ("capture" and "reporter" probes), at least one of which must be specific for the organism to be detected. The capture probe molecules are extended enzymatically with deoxyadenosine (dA) residues at their 3' termini. The reporter probe molecules are labeled chemically with the hapten fluorescein, allowing detection of the target probe complex with rabbit anti-flourescein antibody:horseradish peroxidase enzyme conjugate.

These steps are depicted in Fig. 1, and detailed below. Following cultivation/enrichment, a test sample is transferred in a small aliquot to a test tube and the organisms are lysed, releasing rRNA targets (Step 1). Capture and detector probes are added, and hybridization is allowed to proceed under defined conditions (Step 2). If target rRNA is present in the sample, hybridization takes place between the probes and the target. The solution containing the target/probe complex is then brought into contact with a solid surface (dipstick) containing bound deoxythymidine homopolymer (poly-dT), under conditions that will allow hybridization between the poly-dA and poly-dT (Step 3). Unhybridized nucleic acids and cellular debris are washed away, leaving the captured DNA-RNA complex attached to the surface in via the dA-dT duplex (Step 4). Both reporter probe and capture probe hybridize specifically on the same 16S rRNA target molecule. The entire complex is bound to the plastic dipstick surface via the following interactions: **Capture surface-dT:dA-Capture Probe:Target:Reporter Probe**. The bound fluoresceinated reporter probe is then detected by the addition of an anti-fluorescein antibody:enzyme complex (Step 5).

Following incubation under conditions permitting specific binding of the detection complex and washing to remove non-bound enzyme (Step 6), chromogenic substrate is added, and color development occurs (Step 7). Color development is topped, and the developed color is measured with a spectrophotometer. This reading is compared to the negative control and previously determined cutoff levels, and determination of positive or negative test results is made.

Experimental

Probe development
Sequence information on organism-specific regions of variable domains in the 16S ribosomal RNA (rRNA) of the type strains of specific organisms, i.e. *E. coli, Salmonella, Listeria,* and *Yersinia,* etc. was obtained using the reverse transcriptase dideoxynucleotide termination method of Lane et al., (1985). Specificity of organism-specific probes was predicted by comparison to rRNA sequence available in a number of proprietary and public databases. Complementary DNA oligomers were synthesized and their specificity was determined in dot-blot hybridizations with appropriate panels of gram-negative and gram-positive organisms, according to standard methods described previously (Klinger et al., 1988).

Cultural Methods
Food samples were examined for the presence of *Salmonella* using the procedures specified in the Bacteriological Analytical Manual (AOAC, 1987). Briefly, the culture steps preceding the performance of the assay for *Salmonella* were as follows. One ml. aliquots were transferred from the pre-enrichment culture to selective enrichment broths (selenite cystine [SC] and tetrathionate [TT], as described for the BAM/AOAC method). Selective enrichments to be used for hybridization assays were incubated 6 h at 35°C for all samples except raw meats and raw milk, which were incubated 16-18 h. Following incubation, selective enrichments were vortex-mixed and 1 ml. portions were transferrred from each selective enrichment to separate tubes of GN broth (10 ml./tube). The GN broths were incubated 12-18 h at 35°C with the exception of raw meats and raw milk products which were incubated 6 h prior to assay. The selective enrichments [TT] & [SC] were returned to 35°C for incubation up to total of 24 ± 2 h. 0.25 ml. aliquots from the

STEP 1. SAMPLE LYSIS

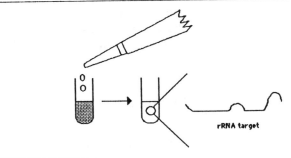

STEP 2. HYBRIDIZATION

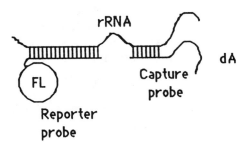

STEP 3. CAPTURE

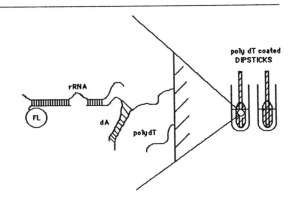

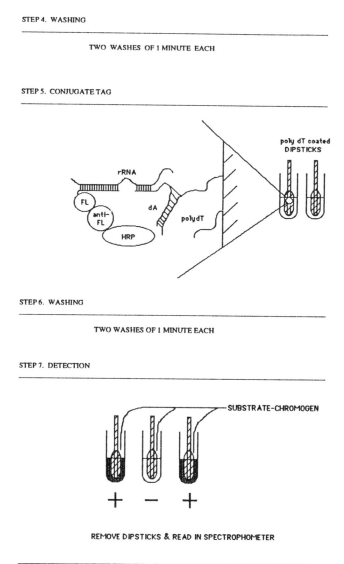

Figure 1. Principles of colorimetric DNA hybridization assay

two GN cultures associated with each sample were combined into a test tube and the assay was performed on 0.5 ml. as described below. The remaining GN and selective enrichment cultures were held at room temperature (20-25°C) until completion of the assay.

Colorimetric DNA hybridization method
GN broth cultures (0.25 ml. from each of the two GN broth per sample) were added to 12x75 mm borosilicate glass tubes. Bacteria were lysed for 5 min. at room temperature by additon of 0.1 ml. 0.75 N NaOH. Samples were neutralized with 0.1 ml. 2M TRIS (pH 7.5); tubes were covered with aluminum foil, and incubated in a 65°C water-bath for 15 min. One-tenth ml. of probe solution was added into each tube, mixed briefly, and returned to the water-bath. Following 15 min. of incubation at 65°C, a poly-dT coated dipstick was placed into each sample tube. Tubes were shaken and returned to the water-bath for an additional 60 min. of incubation at 65°C. During this incubation, 0.75 ml. aliquots of diluted conjugate were transferred to a set of clean 12x75 mm test tubes. After hybridization, dipsticks were removed, blotted onto absorbent paper, and washed for 1 min. at 65°C in pre-warmed 1X wash buffer. Dipsticks were then washed a second time for 1 min. at room temperature in 1X wash buffer. They were removed from the wash basin, blotted, transferred to the tubes containing conjugate, and incubated at room temperature for 20 min. During the 20 min. incubation, 0.75 ml. portions of mixed substrate/chromogen were pipetted into another set of 12x75 mm test tubes. At the end of the conjugate incubation, the dipsticks were removed, washed 2x for 1 min. each in 1X wash buffer, blotted, transferred into tubes containing substrate and chromogen, and incubated at room temperature for 20 min. to allow color development. At the end of this incubation, the dipsticks were removed and discarded, the color development process was terminated with the addition of 0.25 ml. of 4N sulfuric acid. Optical densities (OD) were determined with a spectrophotometer at 450 nm. An OD greater than 0.10 was considered a positive reaction. The positive cutoff value was previously determined by taking the average of at least 30 accumulated negative controls and adding 4 standard deviations.

Samples producing positive hybridization assays were further analyzed to confirm the presence of *Salmonella* by streaking corresponding GN broth cultures to xylose lysine desoxycholate (XLD), Hektoen enteric (HE), and bismuth sulfite (BS) agar plates. Incubation and all subsequent steps in identification of suspicious colonies were performed as described above for the BAM/AOAC methods.

Comparative studies

The GENE-TRAK Colorimetric Assays for *Salmonella* and *E. coli* were compared with the BAM/AOAC culture methods in uninoculated and inoculated foods. For *Salmonella*, samples of twenty (20) food types were tested, i.e., pork, roast beef, turkey, peanut butter, shrimp, coconut, pecans, fish, non-fat dry milk, milk chocolate, pepper, dry casein, gelatin, dry egg, yeast, soy flour, dry pasta, meat & bone meal, cheese powder, and cake mix. Eighteen serovars of *Salmonella* recognized as being frequent isolates from foods were selected for the study (Flowers et al., 1987). Challenge serovars included: *Salmonella anatum, Salmonella agona, Salmonella cerro, Salmonella bovismorbificans, Salmonella cubana, Salmonella derby, Salmonella drypool, Salmonella havana, Salmonella mbandaka, Salmonella meleagridis, Salmonella montevideo, Salmonella new brunswick, Salmonella rubislaw, Salmonella senftenberg, Salmonella typhimurium, Salmonella weltevreden.* These strains used were obtained from various sources, including: American Type Culture Collection, Rockville, MD; Centers for Disease Control, Atlanta, GA; Center for Laboratories, State Laboratory Institute, Jamaica Plain, MA; Silliker Laboratories, Chicago Heights, IL; and University of Massachusetts, Amherst, MA.

Each *Salmonella* serovar was inoculated separately into a sample of the appropriate food type. A cell count was performed on each inoculum broth culture using a serial dilution and plating method and 1-2 ml. of an appropriate dilution was added to the test food to give two desired levels of inocula: low (approximately 0.04-0.2 cells/g) and high (approx. 0.4-2.0 cells/g). After addition of inocula, test foods were mixed using a laboratory blender (Stomacher) to distribute cells evenly before analysis.

Table 1 summarizes the protocol for *Salmonella* enrichment and assay.

Results

Salmonella assay: A total of 371 *Salmonella* isolates representing 223 serovars were tested for hybridization and found to be positive both with the proposed *Salmonella*-specific probe in dot-blot assays and in the final GENE-TRAK colorimetric assay.

Table 1. Protocol for *Salmonella* enrichment and assay

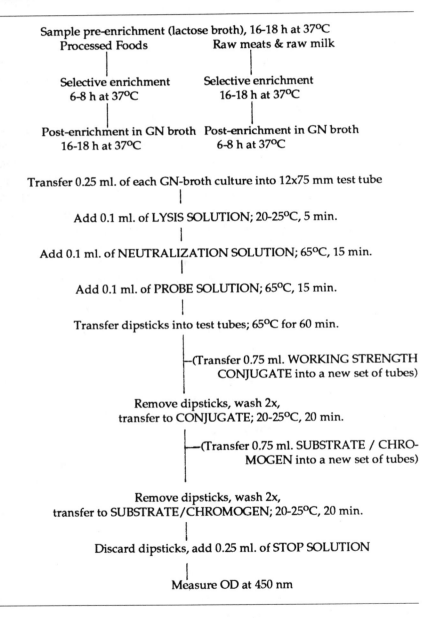

The specificity of the assay was tested further. A total of 107 non-*Salmonella* strains were grown overnight in GN broth and processed as described in the experimental methods. These strains were chosen on the basis of known or suggested physiologic or genetic relatedness to *Salmonella*, or as organisms likely to be found in the flora of food samples. None gave positive results in the assay. The species tested are listed below.

Genus	Species	No. of Strains
Alteromonas	putrefaciens	1
Citrobacter	amalonaticus	3
Citrobacter	diversus	3
Citrobacter	freundii	13
Citrobacter	sp.	9
Enterobacter	aerogenes	3
Enterobacter	agglomerans	13
Enterobacter	amnigenus	1
Enterobacter	sp. CDC grp.19	1
Enterobacter	cloacae	9
Enterbacter	intermedium	1
Enterobacter	gergoviae	1
Enterobacter	sakazakii	2
Enterobacter	taylorae	1
Escherichia	coli	9
Hafnia	alvei	2
Klebsiella	oxytoca	5
Klebsiella	ozaenae	1
Klebsiella	planticola	1
Klebsiella	pneumoniae	6
Klebsiella	terrigena	1
Morganella	morganii	2
Proteus	mirabilis	5
Proteus	vulgaris	2
Serratia	spp.	3
Shigella	boydii	3
Shigella	dysenteriae	1
Shieglla	flexneri	1
Shiegella	sonnei	1
Yersinia	enterocolitica	3

To test sensitivity and specificity, food samples were inoculated with varying levels of each of the 18 *Salmonella* serovars and tested both in the new hybridization assay and by conventional culture (Table 2).

Table 2. Colorimetric hybridization aasay of *Salmonella* in foods: in-house comparative study

Food Type	No. Samples	No. Inoc.	Total Positive[a]	Hybridization Positive[b] (%)	Culture Confirmed Hyb. Pos.[c] (%)	Reference BAM/AOAC Positive (%)
Raw pork	30	24	24	91.7	100	100
Raw turkey	30	24	24	100	100	100
Roast Beef	30	24	24	100	100	100
Raw shrimp	30	24	24	100	92	96
Frozen fish	30	24	24	100	100	95.8
Bone Meal	30	24	26	100	100	100
Gelatin	30	24	24	100	100	100
Dry egg white	30	24	24	100	100	100
Nonfat dry milk	30	24	24	100	100	100
Casein	30	24	24	91.7	100	100
Cheese powder	30	24	24	95.8	100	100
Chocolate	30	24	24	100	100	100
Peanut butter	30	24	16	100	100	100
Pecans	30	24	24	100	100	100
Coconut	30	24	24	100	100	100
Soy flour	30	24	22	100	100	90.9
Cake mix	30	24	24	100	100	100
Pasta	30	24	24	100	100	100
Dry yeast	30	24	20	100	100	100
Black pepper	30	24	24	100	100	100
TOTALS	600	480	469	99.15	98.72	99.36

[a] Total positive by BAM culture method plus additional samples positive by GENE-TRAK Assay and confirmed by plating from GN broth cultures.
[b] Percentage of hybridization positive samples not considering subsequent culture confirmation.
[c] Percentage of hybridization positive samples confirmed by plating from GN broth culture.

Salmonella was detected by either method with nearly equal efficiency. 591 of 600 inoculated or uninoculated samples gave identical results by both procedures. Eleven of the 480 inoculated samples were negative by both hybridization and reference culture methods. The total positive sample population was therefore 469 and the total negative sample population was 131 (including 120 uninoculated samples). There were 6 hybridization false negatives among 480 inoculated samples and 2 hybridization false positive among 120 uninoculated samples. Hybridization false positive and false negative rates were 1.5% and 1.3%, respectively. These results suggest that the hybridization method is equivalent to the current culture method for the detection of *Salmonella* in foods.

E. coli *assay*

A total of 233 *E. coli* isolates were tested for hybridization and found to be positive with *E. coli*-specific probes in both dot-blot assays and in the final colorimetric hybridization format as described above.

The specificity of the assay was tested by performing the *E. coli* hybridization assay with a panel of 163 non-*E. coli* strains. These strains were chosen on the same basis as described in the *Salmonella* section. Except for all *Shigella spp.*, none gave positive results in the assays. The species tested are listed below.

Genus	No. Species	No. Strains	No. Assay Positive
Acinetobacter	1	2	0
Aeromonas	2	2	0
Alcaligene	1	1	0
Bacillus	2	2	0
Candida	2	2	0
Citrobacter	2	15	0
Edwardsiella	1	1	0
Enterobacter	7	25	0
Hafnia	1	1	0
Klebsiella	6	14	0
Listeria	4	9	0
Micrococcus	1	1	0
Morganella	2	2	0
Pasteurella	2	2	0
Proteus	5	9	0
Providencia	4	5	0
Pseudomonas	3	3	0
Salmonella	15	17	0
Serratia	2	2	0
Shigella	4	15	15
Staphlyococcus	3	5	0
Streptococcus	8	8	0
Yersinia	7	14	0

Table 3 shows the results of initial studies using the same colorimetric format for detecting foodborne *E. coli*. In this in-house comparative study, 20 foods were artificially inoculated using one of 16 strains of *E. coli* (including 6 normal fecal isolates, 5 food isolates, 2 enterohemorrhagic [0157:H7] strains, 2 toxigenic, and 1 invasive strain of *E. coli*) at various levels using similar procedures as described in the comparative study of *Salmonella* tests. Briefly, the culture steps preceding the performance of the colorimetric assay for *E. coli* were as follows: Three ml. aliquots were transferred from the pre-enrichment cultures to selective enrichment broth (lauryl sulphate tryptose [LST]) and incubated 24 ± 4 h at 35°C, as described for the BAM/AOAC method with the following modifications. After this primary enrichment, 0.1 ml. of the primary culture was transferred to 10 ml. LST broth (EC broth specified in the Bacteriological Analytical Manual (AOAC, 1984) and incubated another 24 ± 4 h at 35°C (45.5°C recommended by BAM/AOAC).

Table 3. Colorimetric hybridization assay of *E. coli* in foods: in-house comparative study

Food Category	Number Samples	Total Positive[a]	Hybridization Positive[b] (%)	Culture Confirmed Hyb. Pos.[c] (%)	Reference BAM/AOAC Positive (%)
Meats	123	100	99	100	85
Eggs	26	19	94.7	100	78.9
Dairy	65	41	97.6	100	85
Confections	16	10	100	100	80
Grains/Nuts	38	23	100	100	78.3
Fruits/Vegs.	35	21	100	100	71.4
TOTALS	303	214	98.6	100	76.2

[a]Total Positive by BAM culture method plus additional samples positive by GENE-TRAK Assay and confirmed by plating from LST broth cultures.
[b]Percentage of hybridization positive samples not considering of subsequent culture confirmation.
[c]Percentage of hybridization positive samples confirmed by plating from LST broth cultures.

From this secondary enrichment, the *E. coli* colorimetric assay was performed essentially as described above for *Salmonella* except that *E. coli*-specific probes were used, and washes were at room temperature instead of 65°C. Samples producing positive hybridization results were further analyzed to confirm the presence of *E. coli* by streaking the secondary LST broth culture to Levine's eosin methylene blue (L-EMB) agar plates. Incubation and all subsequent steps in identification of suspect colonies were performed as described above for the BAM/AOAC method. The results are summarized in Table 3.

These data indicate superior performance of the hybridization assay over the current cultural method for detection of *E. coli* in foods. There were 3 hybridization false negatives among 214 total positive samples and no hybridization false positives among 89 uninoculated samples. On the other hand, only 163 out of 214 total positive samples were detected using the reference BAM/AOAC method, resulting in a false-negative rate of 23.8%. The reasons for the high false-negative rate for the reference culture method are not clear. Although lactose-negative *E. coli* strains were purposely not chosen for this study, it is still possible that some of the culture misses can be attributed to lack of typical greenish metallic sheen colonies on L-EMB screening due to interference or overgrowth by competitors (unpublished observation).

In the initial screening studies all 15 strains of *Shigella* (representing all four species) were detected using the *E. coli* probes. This was not surprising due to the genetic virtual identity between *E. coli* and *Shigella*. We found that *Shigella* grew reasonably well in the LST selective enrichment and concluded that if *Shigella* was present in a food sample it would be detected in the hybridization assay (data not shown). Certainly the presence of *Shigella* in foods is an important observation, though the incidence is significantly less than, e.g., *E. coli* or *Salmonella* (Smith, 1987). In this instance, however, precise definition of a positive hybridization result would require subsequent biochemical testing.

Finally, we have developed a rapid assay for *Listeria spp.* in the same format. Like the tests for *Salmonella* and *E. coli*, hybridization assays are performed following two days of cultural enrichment. Preliminary results show excellent comparison with both standard culture methods and our current isotopic assay (Klinger et al., 1988). These results will be presented elsewhere.

Discussion

In this paper we have presented data using a dipstick system in non-isotopic hybridization assays for two major foodborne microbial contaminants. These assays will soon enter extensive carefully-controlled (AOAC) collaborative studies with the goal of obtaining official approval (AOAC) as accepted rapid methods. The early field results are very encouraging and suggest that the new assays yield information which is at least as accurate and precise as conventional microbiological procedures.

Nucleic acid hybridization assays in food microbiology have perhaps been compared most often to ELISA products as alternative rapid methods. The assays we have described here rely on probes directed against organism-specific regions of rRNA. Growing data bases of rRNA sequence information from a large variety of organisms, combined with innovative sequencing methods (Lane et al., 1985) have brought new rationality and efficiency to the process of probe selection and development. Additionally, the ability to chemically produce oligonucleotide probes using nucleic acid synthesizers allows reproducible manufacture and quality contol on a highly efficient commercial basis. Certainly monoclonal anitbody techniques have brought similar improvements to the development and performance of immunologically-based assays. However, difficulties in antibody-based assays are perhaps best exemplified in the case of *Salmonella*. A myeloma protein widely reactive with a broad spectrum of *Salmonella* serovars has been found to miss a small percentage of *Salmonella* strains (Mattingly, 1984). When the assay was supplemented with the addition of anti-flagellar monoclonals, performance was increased substantially, but false-positives with apparently antigenically related *Citrobacter* strains continued to be a problem (Mattingly, 1984; D'Aoust & Sewell, 1988). It is of interest not only to diagnostic microbiologists, but also to molecular taxonomists that the probes in our new *Salmonella* hybridization assay clearly distinguish all *Salmonella* from all *Citrobacter* strains tested, under properly controlled conditions of stringency.

While hybridization assays have great potential, and certainly in the dipstick format are simple and adaptable to a variety of test organisms, there is still plenty of room for improvement and innovation. It is beyond the scope of this paper to discuss details of probe development, but even using the rRNA target approach, probes for all organisms are not obtained with equal probability or ease. A key example of such a problem is differentiation of *Shigella* from *E. coli*. These groups share on the order of at least 95% DNA homology and though differences at the rRNA sequence level are demonstrable, they are infrequent. Another challenge area for current hybridization assay procedures in food microbiology and quality control procedures is quantitation. A semi-quantitative result from the enrichment broth tested is a practical possibility. However, because of the sensitivity requirements (1 cell/25 g food), for the time-being a cultural enrichment step between raw sample and material tested will be required. One of the great distinguishing potentials of hybridization methods is amplification of either target nucleic acids, or signal, or both. Target amplification of approximately 10^6-fold in a few hours by automated chemical means has already been demonstrated for a number of pathogens in clinical samples (Saiki et al., 1988). Such amplification approaches may be restricted by problems of fidelity of product amplified, unfavorable background signals, or kinetic problems of interaction of amplification primer probes with the small number of target molecules present in unenriched samples. In theory, however, amplification should allow direct hybridization assays on raw samples containing perhaps as few as one organism.

The rapid assays which we have developed are often used to screen out negatives allowing timely product release and avoiding laboratory expense of working-up and discarding presumptive false-positive samples. As data bases comparing hybridization results with reference standard methods grow, the need for biochemical confirmation of positive results should drastically decline, providing that positive and negative predictive values are adequate. As the menu of the probe sequences expands, a potential advantage for hybridization assays in food microbiology will be the ability to detect other organisms that are difficult or cumbersome to identify. Examples might include *Vibrio* species, spoilage organisms, and various fungi. Combined with amplification, hybridization should open the possibility of routine analysis of foods and water samples for pathogenic viruses. Finally, the FDA and others have recently emphasized the need for assays to identify pathogenic strains. Current procedures such as cell culture, animal challenges, etc., simply do not lend themselves to routine use. In several instances (e.g., *E. coli* and *Staphylococcus* enterotoxins, *Listeria* hemolysins, *Yersinia* virulence plasmids). Nucleic acid sequence information for genes encoding virulence properties or associated with pathogenic serotypes (e.g., *E. coli* 0157:H7) is already available and should facilitate development of hybridization assays.

The assays discussed in this paper were designed as first-generation manual procedures. As information has accumulated about the chemistry of hybridization, hybrid capture, background reduction, and detection, it has become increasingly apparent that these processes can potentially be automated. In the meantime, development of simple, sensitive, colorimetric hybridization assays for this initial group of common foodborne pathogens represents a considerable technical breakthrough. Assuming that they maintain their level of performance as they undergo collaborative evaluation, and that food microbiologists will adopt new procedures if shown to be superior and cost-effective, hybridization assays should have a major impact on providing a safer food supply.

References

A.O.A.C. 1984. Bacteriological Analytical Manual, 6th ed. Assoc. Off. Anal. Chem., Arlington, VA.

A.O.A.C. 1987. Bacteriological Analytical Manual (suppl.) 6th ed. Assoc. Off. Anal. Chem., Arlington, VA.

D'Aoust, J.Y. and Sewell, A.M. 1988. Detection of *Salmonella* with the Bio-Enzabead^{TM} enzyme immunoassay technique. J. Food Protection, 51:538.

Fitts, R. 1985. Development of a DNA-DNA hybridization test for the presence of *Salmonella* in foods. Food Technol. 39:95.

Flowers, R.S., Mozola, M.A., Curiale, M.S., Gabis, D.A. and Silliker, J.H. 1987. Comparative study of a DNA hybridization method and the conventional culture procedure for detection of *Salmonella* in foods. J. Food Sci. 52:781.

Klinger, J.D., Johnson, A., Croan, D., Flynn, P., Whippe, K., Kimball, M., Lawrie, J. and Curiale, M. 1988. Comparative studies of nucleic acid hybridization assay for *Listeria* in foods. J. Assoc. Off. Anal. Chem. 71:669.

Lane, D.J., Pace, B., Olsen, G.J., Stahl, D.A., Sogin, M.L. and Pace, N.R. 1985. Proc. Natl. Acad. Sci. USA 82:6955.

Mattingly, J.A. 1984. An enzyme immunoassay for the detection of all *Salmonella* using a combination of myeloma protein and a hybridoma antibody. J. Immunol. Meth. 73:147.

Saiki, R.K., Gelfand, D.H., Stoffel, S., Scharf, S.J., Higuchi, R., Horn, G.T., Mullis, K.B. and Erlich, H.A. 1988. Primer-directed enzymatic amplification of DNA with a thermostable DNA polymerase. Science 239:487.

Smith, J.L. 1987. *Shigella* as a foodborne pathogen. J. Food Protection 50:788.

Use of RFLPs Analysis to Improve Food Quality

Tim Helentjaris
Molecular Biology Group
NPI
417 Wakara Way
Salt Lake City, UT 84108, USA

Restriction fragment length polymorphisms (RFLPs) can be used as genetic markers to improve crop plants with significant advantages over conventional genetic approaches. Sets of these markers, large enough to basically saturate all areas of the genome, have been produced in several economically-important species such as corn, tomato, brassicas, etc., such that any gene analyzed will be linked to at least one of these markers. When this type of analysis is combined with conventional breeding approaches, many different traits can be addressed with improvements that can result in advantages for the seed producer, the farmer, the processor, and the consumer. Of particular interest is the potential of this technology to facilitate manipulation of quantitative traits. Food quality components such as sugar content in sweet corn and soluble solids in processing tomato have been examined using this type of approach and gains in these attributes can be obtained more economically and quickly than with conventional breeding approaches alone. This type of approach should dramatically improve our utilization of existing variation in important traits, particularly from undeveloped germplasm sources.

Introduction

Many, if not most, of the significant improvements in the utilization of plants as food, from the earliest domestication of species to the development of our current high-yielding hybrids, can be classified as having been the result of conventional genetic manipulation and selection or "plant breeding." The ability to cross various genetic materials and derive progeny with improved characteristics contributed from both parents was recognized long ago and utilized to form most of the basic cultivars in use today. With the additional "discovery" of a basic understanding of genetics in the nineteenth century, this approach has become even more scientific and the resulting improvements in yield of many crop species has been spectacular.

Biotechnology has the potential to impact this process in many ways. Foremost in many people's minds is the potential to create new variability not naturally available through genetic engineering. Besides this approach and including somoclonal variation which can also be used to create variability, there are other technologies that can benefit plant improvement through increasing our ability to better utilize existing genetic variation. For many species there is often already sufficient variation for the traits of most economic interest, but conventional plant improvement approaches are quite slow, requiring on average 10 years to develop and bring an improved cultivar into production. Any technology that can improve this process either by increasing our understanding of the underlying mechanisms of the traits of interest or by enhancing our ability to quickly transfer the responsible genes from one genetic background into another could have as much impact on plant improvement as genetic engineering.

Restriction fragment length polymorphisms
While the improvements in our understanding of basic genetics have had some impact on the strategies of plant improvement programs, it has not been as great as one might suppose, due primarily to some inconspicuous limitations of "Mendelian" genetics. The use of extreme phenotype mutants in marker stocks as the principal genetic tool is sufficient for most academic applications but is simply not practical in the analysis of most advanced germplasm. First, these markers by their nature have phenotypes that are easy to score but at the same time would seriously degrade the performance of these advanced cultivars. Few if any of these types of markers would be expected to be naturally found in advanced germplasm, having been removed during the selection process by the breeder. Secondly, very few extreme phenotype markers exist simultaneously in single lines and the expression of one often interferes with the evaluation of another, precluding the detailed analysis of many loci in single isolates. The nature of "Mendelian" genetics often demands that the geneticist spends most of his time creating informative crosses with the markers of interest to examine a particular problem and a very small amount of time to actually examine the resulting data.

Molecular markers have been proposed as a solution to many of the inherent problems of extreme phenotype markers (Tanksley, 1983). The one example utilized most often previously has been isozymes which have been used to analyze many different types of genetic problems of both an academic as well an applied nature (for examples see

Wendel et al., 1986; Rick and Fobes, 1974). Isozymes possess some powerful advantages such as a silent phenotype as well as relatively low cost of analysis but at the same time often suffer from low numbers of informative loci.

Restriction fragment length polymorphisms (RFLPs) also represent a type of molecular genetic marker (Botstein et al., 1980), while though they currently suffer when compared to isozymes in terms of cost of analysis, they often far surpass them in terms of number of informative loci. Depending upon the application, this advantage far outweighs the disadvantage. It was recognized that variation in genetic sequences between individuals, while not actually involved causally in phenotypic variation, could still be used as a genetic "marker" for the causative genes through the process of linkage of blocks of genetic material into larger functional units or chromosomes. Our current technology to best assay this sequence variation is the use of Southern blots to detect the difference in fragment sizes created by digestion of genomic DNA by restriction enzymes. Future technologies may actually utilize other approaches to find much subtler forms of sequence variation that may be more amenable to automation techniques and solve the problem of high cost of analysis.

Development of RFLP technologies in plants
Sequence variation as detected by Southern blotting and subsequent variation was observed by several researchers and it was quickly recognized that this could be utilized as a means to generate a new type of genetic marker as had been done in the human system (Burr et al., 1983; Rivin et al., 1983; Helentjaris et al., 1985). In general, random pieces of unique sequence DNA, derived from either cDNA or genomic libraries, are tested for their utility in detecting differences amongst the set of germplasm of interest to the plant breeder. We have constructed sets of informative clones for many different species such as maize, tomato, and brassica and other researchers have constructed similar sets in these as well as other species such as lettuce, barley, wheat, and arabidopsis.

In some species this process to construct sets containing several hundred markers is now almost trivial. The current conventional genetic map for maize probably contains on the order of a few hundred markers that have been accumulated during the entirety of this century. In about three years at NPI we were able to produce over 400

RFLP markers and taken together with efforts from other research groups, the total for this species along would be well over a thousand, with more being added all the time. Since many of these markers possess very high degrees of informativeness and can be evaluated simultaneously in single individuals, this is quickly approaching the target of a saturated genetic map where any location in the genome is within a few centimorgans of a usable genetic marker.

Interestingly, all species are not equal with this respect and some have proven very refractory. With maize, approximately 95%+ of the unique sequence clones tested positive for their ability to detect an RFLP between several domesticated inbreds with just three restriction enzymes. On the other hand with tomato, the same figure is less than 5% with even more enzymes. This disparity of the informativeness of random clones holds up for a number of species; brassica is quite polymorphic while with dry beans, soybeans, melons, and lettuce it is quite difficult to isolate cloned sequences that reveal polymorphisms amongst domesticated germplasm. The same division of species is also seen with isozymes and there is a rough correlation with the degree of out-crossing in the reproductive strategies of the two classes. Those that are primarily out-crossing in nature reveal a high degree of polymorphism and those that are primarily self- pollinated in nature are relatively monomorphic. The basis underlying this correlation is not clear, although a number of hypotheses are evident.

The collection of cloned sequences that reveal informative RFLP loci for any species can then be used as a tool set in many different applications to problems of both an academic as well as a practical nature. Since relatively large sets can be easily generated, a way was needed to systematically use such a tool set. One could simply select clones at random but this would hardly be efficient with sets greater than 400 loci. The key is that these markers are useful by the nature of their linkage to other loci in the genome. By understanding the linkage relationship of the marker clones with respect to each other, one could efficiently select from the set such that the bulk of the genome could be tested with a moderately small subset of the entire set of clones.

This process is known as constructing a linkage map. To accomplish this, one prepares a population which can be expected to be segregating for most of the loci detected by the RFLP marker clones. Usually an F2 population is derived by crossing two divergent inbreds to create a heterozygous F1 population which is then selfed to produce the segregating F2. The marker clones are then tested for their genotypes in a set of F2 individuals, usually on the order of 100-200, and the tabulated genotypes for all marker loci are compared. Linkage analysis will then yield which loci are linked together onto the same chromosome, the spatial order of these loci, and finally the recombinational distances between marker loci. In some cases the linkage groups have been identified to described chromosomes and in others with less developed genetics, they remain as simply linkage groups. Linkage maps based upon RFLPs have now been produced for a number of species including maize, tomato, lettuce, brassica and others (Helentjaris et al., 1986; Bernatsky and Tanksley 1986, Landry et al., 1987, M. Slocum unpublished results).

Application of RFLPs to problems in plant improvement
The construction of the linkage map of a set of informative clones can be thought of as the construction of a tool set. This set can prove to be very useful for a number of applications in plant improvement programs. By testing clones distributed at random throughout the genome against a set of germplasm, one can order that set of germplasm as to the member's relatedness by actual genetic distance. We have done this in maize for instance (Walton and Helentjaris, in press) and find that the relationships determined by use of RFLPs alone are quite accurate when compared with our knowledge of those relationships from the actual pedigrees of the inbreds used. In fact this type of information is much more accurate from the standpoint that one can not only determine that two lines are related at a certain level, but can also determine which chromosomal blocks are similar and which are different.

The significance of this can be seen by observing that for a species like alfalfa, which is usually cultivated as a synthetic of twelve or more isolates, the task facing the breeder of evaluating all possible combinations of twelve isolates from even a small group of 50 possible parental isolates is very impractical. By using RFLP information to order that set of 50 test isolates into groups related by ancestry and probably also function and then selecting representatives from each group for testing, the breeder can reduce the complexity of his testing program by an order of magnitude. The application to a crop like soybeans is also important where numerous collections have been made and where little information is available on the relationships between isolates in those collections. RFLP analysis can complement the traditional types of visual observation used to attempt to establish relationships between individuals.

As a further demonstration of the utility of using RFLP information to understand the relationships amongst a set of advance germplasm, we also examined the predictive ability of RFLP-derived genotypes for yield in hybrid species. We found that if we plotted genetic dissimilarity for a number of maize inbred pairs, as determined solely by RFLP analysis, vs. the yield of the hybrid when these inbreds were crossed to produce a F1, there was a striking relationship. The yield was almost linearly related to the dissimilarity of the parental inbreds up to about the 50% level where it leveled off and further increases in dissimilarity had no measurable impact on yield. This result should not be seen as too surprising from the standpoint that similar but cruder relationships were also seen with isozymes and it has been a general understanding amongst breeders that the more unrelated the parental lines, the better chance of producing a successful hybrid. This result also points out a fact made earlier as to the usefulness of biotechnology in impacting plant improvement. One possible outcome of this work is that we may be able to generate a predictive model for hybrids that will allow the breeder to initially select based upon actual genotype instead of inferences from pedigree and intuition. At the very least these types of studies will also significantly increase our understanding of the underlying mechanisms of heterosis and allow us to redesign our conventional breeding programs to be more efficient.

Additional benefits to the plant breeder and the consumer as well are the use of RFLP analysis to "fingerprint" isolates and determine their identity and purity. For the developer it is important to retain his investment in the development of superior genetic materials by preventing their unauthorized use by others. Fingerprinting by RFLPs can provide a much more accurate method of identifying materials than by strictly visual observation as is also the case with the current use of isozymes. The impact can be similar as with human paternity testing where the initial use of blood-typing was sufficient for establishing non-identity but inconclusive for proving identity. RFLP analysis can now provide a statistically significant identification for cultivars in many species akin to that accepted legally for fingerprints in human identification. RFLP analysis can also be used as with isozymes to establish the genetic purity of a mixed sample as is often the case with seed lots. It is important to demonstrate to the consumer that the lot of seed he has purchased as being genetically superior is indeed completely of the expected genotype.

An additional application for RFLP analysis is the use of linked markers to facilitate the transfer of commercially important genes from one genetic background to another. While in the past, the expected useful lifetime of a commercial cultivar could be on the order of twenty years or more, currently with the rapid improvements in many programs, this figure has dropped to less than ten years in many cases. Since as we mentioned earlier it can often require ten years to successfully commercialize a new cultivar, this puts tremendous pressure on the plant breeder to be both fast and efficient in his programs. Yet the reliance previously upon selection based upon phenotype has many serious drawbacks. The expression of a phenotype can be significantly affected by the environment in which the test plants are grown, usually limiting evaluations to once a year in the area directly of interest. With the widespread effects of the drought in the midwest this year, a whole year of progress was lost for many programs. If plants could be selected based upon their genotype, it does not matter where the plant material is prepared and evaluations could be accomplished up to three times a year and with green-house grown materials near a central convenient site. Because of this environmental influence on phenotypic expression it often means that tests must be replicated many times and in several locations and it may take years before the evaluations can be considered to be satisfactory. Direct detection of the genes underlying the beneficial phenotype may mean that much of this replication will become superfluous, dramatically reducing both the time and cost of development of new and improved materials.

Since many genes which are causative for the desirable phenotypes are often recessive in their expression (i.e. they will only be evident in the homozygous state in a diploid individual), their introgression from one genotype to another is very slow requiring several additional generations of selfing to homozygosity and evaluation. Evaluation at the genotypic level can determine the presence of "silent" genes in the heterozygous state during a breeding program and remove the need for these very slow additional selfing and evaluation steps.

Our previous model for this application of RFLPs as molecular markers for traits of interest are the use of isozymes. They have been used to map traits of commercial interest previously and in one striking example were used to introgress a nematode resistance factor from an undomesticated tomato relative into commercially useful germplasm and this particular process is now the one of choice for the industry (Rick and Fobes 1974). Isozymes will continue to be very useful due to their low cost of analysis but there are often no useful and informative isozymes located close enough to the gene of interest to be of practical use. The availability of saturated RFLP linkage maps to identify and select a marker in close enough proximity for use in the linked introgression of single gene traits will substantially improve our ability to transfer genes between genotypes. In our own efforts we have been able to link RFLP markers to disease resistance loci in both maize and brassica (Helentjaris et al.; Figdore and Slocum, unpublished data) and anticipate that these linkages will be useful in transferring these traits into more useful genetic backgrounds. Disease resistance factors particularly are often due to single major genes and will benefit from this approach but also many other traits could be manipulated more efficiently such as starch and sugar levels and types in corn and oil level and content in soybeans and oil seed rape.

One of the difficulties in plant improvement is that often the genes of interest lie in undeveloped germplasm. In the case mentioned above with nematode resistance, the source germplasm actually consisted of a related species of our domesticated tomato. Obviously one wants only to transfer the gene of interest and leave behind all of the deleterious factors of this undeveloped material. It has been a consistent problem of how to move the gene of interest into a highly developed cultivar and not destroy the years or decades of progress in improving the other characteristics of that cultivar. One use of RFLPs can be to not only screen the segregating progeny for the presence of the genes of interest, but to also screen individual progeny for the most resemblance to the overall desired parental type. This can mean that instead of relying upon the slow process of back-crossing to statistically reach a certain level of resemblance to the desired type, one might actually screen with RFLPs unlinked to the genes of interest to select progeny that are one or more generations advanced beyond the statistical average for that generation and save several steps of back-crossing. With time being an important factor in any program, this may improve our ability to utilize related but undomesticated isolates available in nature and add justification to the extensive efforts in conservation of plant species around the world. In one effort, we were able to analyze an insect resistance trait in tomato and not only use RFLPs to facilitate the transfer of the causative genes, but also use unlinked RFLPs to identify individual progeny that were at least one generation advanced over that level we would have expected on average for the group as a whole (Nienhuis et al., 1987). This study also pointed out one of the complications of using wide crosses in breeding programs, where usually one can assume that inheritance of the two types of chromosomes from each original parent is essentially equal, we detected strong biases in some cases in favor of the undomesticated parental type. This type of factor could strongly work against any conventional program where the reliance was on statistical inference of the genotype and points out the utility of using marker-based selection.

While the application of RFLPs as linked molecular markers to single gene traits is fairly obvious, a more important application will be the use of RFLP analysis to analyze and dissect multigenic or "quantitative" traits into their individual genetic components. Most of our commercially important traits of interest to the breeder, producer, and consumer are multigenic in nature, factors such as yield, maturity, quality and content of carbohydrates and oils, flavor, etc. The complex genetic basis of these traits taken together with the confounding influence of the environment upon their expression means that these are also the most difficult traits for the plant breeder to manipulate in his programs. Replication is required for both location and years to insure satisfactory evaluation but is usually limited to one test per year.

Again however, work with isozymes as molecular markers has provided a model for the application of RFLPs to this problem. Earlier studies with isozymes in maize showed that segments of the variation for a number of important traits such as yield and ear number could be accounted for by major genes linked to isozyme loci (Stuber, et al., 1982). These studies also suggested that marker-assisted selection could prove more powerful than field evaluations alone. We have taken advantage of the greater numbers of RFLP loci in our maize map to dissect some quantitative traits in greater resolution than was previously possible. By examining such traits as height and yield components in corn, we were able to determine that a number of major genes existed for these factors and that taken together, one could explain a level of variation for these traits that should prove useful in selective models (Edwards, Stuber, Helentjaris, and Wright, unpublished data). It was found that a number of major genes could account for a significant amount of the phenotypic variation for plant height and also ear height. One goal of corn improvement has been to maintain plant height but lower the height of the ear so that the plant is less susceptible to falling over in the field. Interestingly all of the loci in this cross that accounted for plant height also were involved in ear height, suggesting that separation of these two traits would be virtually impossible. Analysis of this type would then determine very early in such a program that this approach would be futile and that this effort should be discontinued in these genetic materials. Hence direct selection with RFLPs is not even necessary to have a major impact on practical plant improvement.

During these studies we were also able to determine that there were major genes that affected total yield of the progeny and that this yield could be broken down into components with different genetic locations and gene actions. No loci were found to affect yield in this experiment by increasing seed size but all acted by affecting numbers of seeds yielded per plant. This factor could be broken down further into genetic factors that increased seed yield per plant by increasing the number of ears per plant and others that affected ear morphology (girth or length) such that more kernels were produced per ear. Hence one does not need to manipulate this trait blindly but could select for plants with multiple ears to improve yield or for different production reasons, plants with larger single ears. One can see that this can lead to a very directed program to improve even complex traits such as yield by breaking them down into components, determining the loci most responsible for the expression of those components, and then using marker-assisted selection to bring together the most productive combination of components.

In another example even more pertinent to food quality, we have used RFLPs to analyze and dissect the trait of soluble solids in processing tomatoes (Nienhuis et al., unpublished results). Yield in processing tomatoes is based upon soluble solids as one of its major components and small increments in improvement of the total soluble solids are very valuable to the processor who is willing to pass some of that along to the grower if he can supply improved product. It had been previously identified that a wild species, *L. chmielewskii*, possessed much higher levels of solids, >10% vs 5% for most commercial cultivars, and some attempts had been made to transfer some components of this trait into a commercially acceptable cultivar. However besides the complex genetic nature of this trait, this species also possesses many deleterious factors, such as very small fruit, low yield, indeterminate growth character, and green fruit color, all of which were unacceptable to the grower, processor, and consumer. Consequently little progress had been made in improving this particular trait despite the major importance attached to it. We used RFLP analysis to dissect this trait and found several major loci that, taken together in a predictive model, meant that we could account for over 50% of the phenotypic variation for this particular trait. We tested this predictive model in lab and field selections and demonstrated that one could select based upon RFLP analysis and make significant progress over conventional breeding approaches. The importance of this can also be seen in that the actual phenotypic evaluation of this trait is very difficult and occupied a whole crew for one summer to test 100 hundred isolates. With the predictive value of the RFLPs, it was rather trivial for a single lab technician to screen material in the greenhouse at the level of several hundred individuals in the off season and select the top fraction of promising individuals for further evaluation in the field next season. Not only did we significantly impact the efficiency of selection for this trait, but other markers were also found that were linked to other important factors such as yield and fruit color, so that simultaneous selection for several important traits could be effected. The immediate result of this program is that commercially-acceptable hybrid combinations were identified that were greater than 7% in soluble solids which could be worth several millions to the processor in improved product and savings in operation costs. Even more importantly using the RFLP analysis we know that none of the improved progeny contain all of the important loci in the optimal combination, and that efforts in a very directed manner can continue to produce this optimal combination by utilizing only those progeny predicted to be useful, a process which would be impossible by strictly conventional approaches.

Future directions for this technology
Clearly the potential of many approaches that will impact plant improvement programs have been demonstrated and work continues in many groups to actually bring these to fruition. The major handicap at this time is cost and complexity of analysis which practically limits this to only the most important applications or those that are impossible by conventional approaches. As with many innovations, the simple

demonstration of utility then forces the work in the direction of practical implementation and many groups are now involved in attempting to modify the technology to allow far greater throughput, on the order of tens of thousands of samples per year for single companies, at less cost and in less time. The current technology was sufficient for most academic applications but is very cumbersome and difficult to envision automating to the degree required. The realization that Southern blotting is not the only method to detect the underlying sequence variation means that quite different technologies may evolve which will be more amenable to automation. The ability to detect single base pair alterations may also mean that we can extend this technology into other species previously considered to be refractory to this approach.

It is anticipated that when this happens, the advantages of genotypic selection will make this process a major part of most plant improvement programs, replacing to a great degree the early stages of field evaluation which are most difficult for reliable measurement. The ability to address quantitative traits to a degree which is now unenvisioned should also mean that we may see a great jump in the progress made in those characteristics which on one hand are considered the most important but on the other, also considered the most difficult.

All of these applications have essentially dealt with improving our conventional plant improvement strategies but the promise is also there to substantially affect our attempts at genetic engineering of commercial cultivars. Currently we are limited to attempting to engineer simple traits such as insect resistance where the source of the gene is usually not a plant at all. Our inability to engineer the commercially important traits such as yield is due to a basic lack of understanding of the mechanism underlying these complex traits. Use of RFLP analysis should substantially increase this understanding and begin to allow us to consider how one might alter a trait which involves the expression of large numbers of genes. In one example we have found that some of the loci involved in complex traits in maize as mapped through RFLP analysis are located near described extreme phenotype genes (Edwards et al., unpublished data). In one example a quantitative trait locus (QTL) for both height and yield was found to be located very closely to a known plant hormone deficient dwarf on chromosome 9. Other similarities between the extreme phenotype and the QTL were also noted. If this relationship could be confirmed, it would facilitate the cloning of the locus, its *in vitro* engineering, and reintroduction to perhaps affect a complex and economically important phenotype.

It may turn out that only a few or even one gene needs to be altered to obtain much greater than natural variation for a relatively complex trait. In a model worth studying further (Palmiter et al., 1983), researchers found that if they engineered the growth hormone gene with a higher level promoter and reintroduced it back into mice, they obtained some progeny whose body size was tremendously increased, much beyond what has been seen naturally. Body size is obviously a very complex trait and yet the alteration of a single gene resulted in a tremendous effect upon its expression. Even though a trait may be realized to be the result of the expression of many genes, alteration of a few key or "rate-limiting" gene products may be able to result in the generation of tremendous variation, some of which may be commercially useful. Besides the impact of RFLP analysis on conventional plant improvement in the short term, it may also greatly impact genetic engineering by helping us to pinpoint rate-limiting genes whose alteration will be crucial to affecting trait expression as well actually facilitate cloning of those loci.

References

Bernatsky, R. and Tanksley, S. 1986. Toward a saturated linkage map in tomato based on isozymes and random cDNA sequences. Genetics 112: 887-898.

Botstein, D., White, R., Skolnick, M. and Davis, R. W. 1980. Construction of a genetic linkage map in man using restriction fragment length polymorphisms. Am. J. Hum. Genet. 32: 314-331.

Burr, B., Evola, E., Burr, F. A. and Beckmann, J. S. 1983. the application of restriction fragment length polymorphisms to plant improvement. In: "Genetic Engineering Principles and Methods", Vol. 5, pp. 45-59. Plenum Press, New York and London.

Helentjaris, T., King, G., Slocum, M., Siedenstrang, C. and Wegman, S. 1985. Restriction fragment polymorphisms as probes for plant diversity and their development as tools for applied plant breeding. Pl. Mol. Biol. 5: 109-118.

Helentjaris, T., Slocum, M., Wright, S., Schaefer, A. and Nienhuis, J. 1986. Construction of genetic linkage maps in maize and tomato using restriction fragment length polymorphisms. Theor. Appl. Genet. 72: 761-769.

Landry, B. S., Kesseli, R. V., Farrara, B. and Michelmore, R. W. A genetic map of lettuce (*Lactuca sativa* L.) with restriction fragment length polymorphism, isozyme, disease resistance and morphological markers. Genetics 116: 331-337.

Nienhuis, J., Helentjaris, T., Slocum, M., Ruggero, B. and Schaefer, A. 1987. Restriction fragment length polymorphism analysis of loci associated with insect resistance in tomato. Crop Sci. 27: 797-803.

Palmiter, R. D., Norstedt, G., Gelinas, R. E., Hammer, R. E. and Brinster, R. L. 1983. Metallothionein-human GH fusion genes stimulate growth of mice. Science 222: 809-814.

Rick, C. M. and Fobes, J. F. 1974. Association of an allozyme with nematode resistance. Rept. Tomato Genet. coop. 24: 25.

Rivin, C. J., Zimmer, E. A., Cullis, C. A., Walbot, V., Huynh, T. and Davis, R. W. 1983. Evaluation of genomic variability at the nucleic acid level. Pl. Mol. Biol. Rep. 1: 9-16.

Stuber, C. W., Goodman, M. M. and Moll, R. H. 1982. Improvement of yield and ear number resulting from selection at allozyme loci in a maize population. Crop Sci. 22: 737-740.

Tanksley, S. 1983. Molecular markers in plant breeding. Pl. Mol. Biol. Rep. 1: 3-8.

Walton, M. and Helentjaris, T. in press. Application of restriction fragment length polymorphisms (RFLP) technology to maize breeding. Amer. Seed Trade Assoc.

Wendel, J. F., Stuber, C. W., Edwards, M. D. and Goodman, M. M. 1986. Duplicated chromosome segments in *Zea may* L.: further evidence from Hexokinase isozymes. Theor. Appl. Genet. 72: 178-185.

Hybridoma Technology: The Golden Age and Beyond

R.A. Goldsby
Department of Veterinary & Animal Science
University of Massachusetts at Amherst
Amherst, MA 01003, USA

The production and use of monoclonal antibodies is now as routine as the isolation and use of enzymes. Although the basic elements of deriving and producing monoclonal antibodies have remained the same as those established by Kohler and Milstein, important and useful modifications have been introduced. Gradually, strategies have been worked out which permit the production of monoclonal antibodies from humans and animals of veterinary interest. Technical modifications such as *in vitro* immunization and partner selection have extended significantly the power of the basic method. The evolution of hybridoma technology continues to provide greater capacity to engineer and "customize" monoclonal antibodies. Representative of this trend is the construction of chimeric molecules in which investigator-selected domains of different antibodies, or even enzymes, are mixed and matched. The next developments in this field will likely involve a continuing fusion of such technologies as cell hybridization, recombinant DNA methodology and with the recent discovery of catalytic antibodies, enzymology.

Introduction

In 1975, the publication of Kohler and Milstein's now classic paper, "Continuous cultures of fused cells secreting antibody of predefined specificity", changed the practice of pure and applied immunology. The ability of hybridoma technology to isolate and indefinitely amplify the monoclonal immunoglobulin producing potential of individual B cells has allowed immunologists to pose and answer a broad spectrum of fundamental questions. On the practical side, monoclonal antibodies have found uses in the areas of quality control, purification, diagnosis and experimental therapies. This still widening spectrum of applications has spawned and supports a lively sector of the biotechnology industry.

From its inception, the derivation of monoclonal antibodies has been one of the most accessible of laboratory technologies. With a modest capital investment, the application of simple immunological methods and competent cell culture technique, almost any determined laboratory can produce useful reagents. The essential conception underlying the construction of antibody-secreting hybridomas has remained unchanged in the years since 1975. First, one immunizes the system with the antigen of interest in order to predefine the specificity of the prospective hybridomas. Second, cells from the immunized system are fused to a cell line which is permissive for immunoglobulin production. Third, the antibody products of hybridoma clones are screened to identify those antibodies possessing the desired characteristics. Finally, the hybridomas that secrete antibody with the desired characteristics are grown in mass culture and the supernatants harvested, or, more economically, as tumors whose antibody product appears in milligram quantities in serum and ascitic fluids. However, over the years, each of these steps has undergone significant technical evolution. The end result of these advances, some of which will be noted in this review, has been a useful expansion in the range of the technology. Also, there has been a significant increase in the opportunity to control such variables as culture conditions and scale-up. It is even possible to select particular B cells in a population of immunocytes for preferential fusion on the basis of their antigen specificity or even the affinity of their surface immunoglobulin. Furthermore, the fusion of hybridoma-based methodology with other powerful approaches, such as recombinant DNA technology and lately, enzymology, has brought a scope and power to the monoclonal approach barely glimpsed or completely unforeseen in the early days of its development. This review, an update of one I presented just a few years ago (Goldsby, 1986) will provide an appreciation of the current state of the field and suggest some of the likely pathways of future development.

Making Hybrids: Today and Tomorrow

The march of technology has brought significant improvements to the derivation and growth of antibody secreting hybridomas. These include *in vitro* immunization, the use species other than the mouse as a source of normal lymphocytes, the possibility of using partner selection techniques and electrofusion as an alternative to polyethylene glycol (PEG). However, for now, and the near future, the overwhelming majority of useful monoclonal antibodies will be made by the PEG-assisted fusion of spleen cells from conventionally immunized mice with an appropriate established cell line.

With respect to the use of species other than the mouse, one notes that antibody-secreting hybrids have been obtained by fusing a myeloma derived established cell line with rat (Galfre et al., 1979), human (Olsson & Kaplan 1980) bovine (Srikumaran et al., 1983), porcine (Raybould et al., 1985) and recently, rabbit (Raybould and Takahash, 1988) lymphocytes. Since myeloma cell lines have been established in only a relatively few species, it is often necessary to hybridize across species lines. Thus when lymphocytes are used from most of the non-murine species mentioned above, investigators have had to stabilize interspecific hybridomas in which the established cell partner is of murine origin. This has been complicated by the problem of greater instability of karyotype which is the inevitable consequence of deriving hybrids across species lines. However, many laboratories have managed to live with this problem by diligently cloning and recloning secreting cell populations in order to derive and maintain sublines. In some cases the sublines derived by this strategy have secretory phenotypes as stable as those of their murine X murine counterparts. Also, we should note that there are highly satisfactory rat myeloma lines (Bazin, 1982) one of which, IR983F, is derived from the inbred Lou C/n rat and hence is ideally suited for raising ascities from desirable hybridomas derived from Lou C/n X Lou C/n fusions.

Because of the obvious importance of human monoclonal antibodies, a good deal of effort has gone into the development of myeloma and lymphoblastoid cell lines which will efficiently partner fusion with human lymphocytes to make antibody-secreting human X human hybridomas (reviewed by Thompson, 1988). Although success has been achieved using human cell lines as the established cell partner, the efficiencies of hybridoma recovery are not as high and the level of antibody secretion is often much lower than that seen in murine X murine hybridomas. In order to improve the yield of hybrids, some workers have used a two step procedure to immortalize the secretion of human antibody. The first step involves transformation of donor B lymphocytes with Epstein-Barr virus followed by a fusion of the transformed lymphocyte population with an established human Ig-synthesis EBV-transformation as a preliminary step to immortalize and expand a polyclonal population of human lymphocytes from donors judged suitably immunized, much remains to be done before the production of human X human antibody-secreting hybridomas becomes as routine and facile an exercise as the derivation of counterpart murine hybridomas.

Over the years there have been important advances in the rational design of media for the derivation and culture of hybridomas. Initially, these efforts were driven by the advantages of using culture media in which the serum content was reduced or eliminated. A variety of media formulations were advanced to meet this need. All of these were supplemented with three ingredients, insulin, transferrin and selenium (InTrSe). Using media supplemented with InTrSe, it was possible, using routine procedures, to grow most hybrids at greatly reduced levels of serum (2-5%) and when suitable steps were taken, to derive and grow hybrids in serum-free medium. These suitable steps involved adapting the established cell line to vigorous growth in the supplemented serum-free medium and taking care to avoid growth of hybrids to high densities.

Recently, efforts have reached beyond the obvious advantages for consistency, and ease of downstream processing afforded by the elimination of serum. This work has sought to identify factors which specifically support the maintenance and growth of cells with the immunoglobulin secretory phenotype. During the past few years, a cytokine, IL-6 has been recognized and reported by several groups (reviewed by Sugasawara, 1988). This growth factor which as been recognized in cultures stimulated with murine peritoneal macrophages, and L929 cultures stimulated with double-stranded RNA, has been reported to produce significant enhancements (in excess of 10 fold) in the number of antibody-producing hybridoma colonies obtained after fusion of splenocytes from immunized mice with SP 2/0. We can expect the coupling of the hybridoma growth factor, IL-6 (and possibly others yet to be discovered), with advances in the technology of cell growth in serum-free defined media to be widely adopted. The end result of such improvements will be greater yields of potentially useful hybridomas, better reproducibility in their derivation, significant facilitation of the monoclonal antibodies produced and possible important increases in the yields of antibody secreted.

In vitro Immunization

A few years ago, Luben and Mohler (1982) demonstrated that one could bypass the immunization of animals and simple immunize cultures containing appropriate immunocytes, spleen cells for example, with the antigen against one wished to raise monoclonal antibodies.

The pioneering studies of these investigators demonstrated that *in vitro* immunization offers the following advantages:

- The immunization procedure requires only 5 to 7 days.
- Defined levels of antigen can be maintained throughout the immunization.
- There is the possibility of utilizing regulatory cytokines to attempt modulation of the response; i.e. antibody class.
- Successful immunizations can be accomplished with small amounts of antigen (10-100 nanograms have worked).
- Antigens which are toxic at the organismal level may be innocuous in cell culture (e.g. some neurotoxins).
- It has been possible to produce antibodies to highly conserved antigens which are difficult to produce by conventional immunization because of tolerance.

Surely, *in vitro* technology will continue to evolve in the next few years. We can predict that the improvements in the understanding of the factors critical for the growth of antibody secreting cells will be incorporated into *in vitro* immunization protocols with concomitant increases in the efficiency of the technique. Certainly, considerations of convenience and feasibility will make this a technique even more widely applied in future years to the production of human monoclonal antibodies as well as those from large animal species of veterinary interest.

Partner Selection

As commonly practiced, the production of antibody secreting hybridomas is only a partially defined process. One usually settles for the partnering of the members of an immunocyte preparation from an immunized animal or culture with an Ig-permissive cell line. Consequently, a number, perhaps the majority, of the fusion events take place between cells of no interest to the experimenter and the established cell line. Viewed from the perspective of the post-fusion activities of the experimenter, conventional methods of hybrdoma production inevitably lead to the culture of many hybrid cell lines which are not producing antibody of the desired specificity. Consequently, hybridoma production as usually practiced falls short of the platonic ideal of uniform derivation of "continuous cultures of fused cells secreting [monoclonal] antibody of predefined specificity." This need not be the case. Methods of partner selection have been devised which can replace the shotgun lottery of "stick and pick." Two approaches to partner selection are outlined below.

In the procedure dubbed receptor-mediated electrically induced fusion, myeloma cells and antigen specific B cells are preferentially linked to each other by a biotin-antigen-immunoglobulin bridge (Lo et al., 1984). The specifically partnered cells are then efficiently fused by the application of an electric field. As demonstrated by Zimmerman and his colleagues (Zimmerman and Vienken, 1984). The preferential linkage is ingeniously engineered in the following manner. Biotin is attached to the surface of myeloma cells by reaction of appropriately positioned nucleophilic groups of cell surface macromolecules with N-hydroxysuccinimide-linked biotin. The antigen of interest is cross-linked to avidin by means of 1,5-dinitrofluoro-2, 4-dinitrobenzene. Addition of the avidin-antigen couple to a suspension of spleen cells from a mouse immunized with the antigen results in the specific binding of the couple to the subpopulation of B cells which bear surface immunoglobulin specific for that antigen. When biotinylated myeloma cells are added, there is preferential association with the B cells which have avidin attached via the bound antigen.

In an impressive demonstration of the power of the technique, these investigators derived monoclonal antibodies against angiotensin-converting-enzyme, against enkephalin convertase, and against the inflammatory nonapeptide, bradykinin. An important feature of the hybridomas produced by this procedure was that all of those recovered produced antibody of the desired specificity. Clearly, the use of the appropriate antigen-avidin conjugate can effectively restrict most of the fusions to biotinylated established cell lines and the small subset of B cells specific for the coupled antigen. It is of interest to see if this receptor-mediated focusing technique, teamed with electrofusion by Tsong and his colleagues, can be used with other fusing strategies, such as that chemically mediated by PEG.

A second approach introduced by Parks and his colleagues (Parks, 1979) uses the coupled capacities of the fluorescence-activated cell sorter (FACS) for cell identification and isolation of cells displaying predetermined fluorescence profiles. Those members of a freshly generated hybridoma population that bear immunoglobulin molecules specific for the immunizing antigen will bind molecules of the antigen if they are presented to them. Thus, the addition of antigen which has been conjugated to an appropriate fluorescent label will specifically tag those hybridomas that are synthesizing antibody which is specific for the labeled antigen. Such tagged hybridomas can be sterilely sorted from other members of the popualtion and cloned to yield pure lines of cells secreting the antibody of interest.

Catalytic Antibodies

The fact that the active sites of many enzymes are structurally and electronically complementary to the transition staets of the reactants led Jencks (1969) to suggest that ". . . it should be possible to synthesize in enzyme by constructing a binding site. One way to do this is to prepare an antibody to a haptenic group which resenbles the transition state of a given reaction." One would expect that such an antibody should lower the free energy of activation by stabilization of the transition state and thereby exert an enzyme-like acceleration of the reaction rate. This suggestion, made almost 20 years age, has proven to be prophetic.

During the past two years, two groups have independently demonstrated that antibodies specific for structures resembling the transition states encountered during the hydrolysis of carbonates and esters selectively catalyze the hydrolytic cleavage of their analogous substrates. Specifically, the laboratory of Schultz has derived and characterized antibodies that catalyze the hydrolysis of methyl p-nitrophenyl carbonate (Pollack et al., 1986). The Lerner laboratory (Tramontano et al., 1986) has derived an antibody against a picolinic acid-bearing phosphonate that catalyzes the hydrolysis of an analogous aryl ester which lacks the picolinic moiety. Both of these groups have extended their initial findings (see Schultz, 1988 and Napper et al., 1987).

The current expectations that a variety of antibodies with catalytic activities will be found, are fueled by the recognition that the potential for structural diversity inherent in the immune system is enormous, well in excess of 10^8. There is a high likelihood that such immunoglobulin-based catalysts would find application in the specific modification, hydrolysis or condensation of a variety of molecules of interest to pharmacology, biology and materials chemistry. The potential applications of "abzymes" notwithstanding, there is an important obstacle to the general application of antibodies to the catalysis of a wide variety of different chemical reactions. To date the key to the derivation of a catalytic antibody has been the availability of a haptenic transition state analog which could be coupled to an immunogenic carrier. The haptenated transition state analog is then used to raise analog-specific antibodies. The problems which will keep such an approach from being generally applied are apparent. First, current chemical theory cannot precisely divine the structure of the transition state for many ractions of interest. Secondly, at a practical level, even when the structure of the transition state is known, there is the problem of synthesizing a transition state analog which can be coupled to an immunogenic carrier. A possible alternative to these problems is suggested in the section below.

Specific Antibodies Without Specific Immunization

Polyclonal B cell activators (PBAs) are substances that directly activate B cells to divide and differentiate into immunoglobulin-secreting cells. The stimulation of mouse B lymphocytes by a PBA results in the division of each mitogen-reactive B cell to produce a clone of cells which secretes Ig-specific for antigens of a given configuration. Because cultures stimulated by PBAs are highly polyclonal and each Ig-secreting clone produces Ig of particular specificity, B cell cultures activated by such mitogens contain antibodies with a wide variety of specificities for a broad diversity of antigens. The fusion of such mitogen stimulated cells from unimmunized donors results in the production of hybridomas which secrete antibodies displaying a broad range of specificities (Goldsby et al., 1978; Andersson and Melchers, 1978). Such an approach may be used to construct a repertoire-representative "library" of monoclonal antibody secreting hybridomas. Among the members of this library one might expect to find "catalytic" antibodies and antibodies with a broad diversity of useful specificities. It is important to note that this route to catalytic antibodies requires no knowledge of the structure of transition states and no artful synthesis of analogs for purposes of immunization.

Chimeric Antibodies

Chimeric antibodies are the offspring of a fusion of recombinant DNA technology with hybridoma methodology. They are the protein products of chimeric genes, constructed by joining immunoglobulin V region genes from one species with genes from another species. The transfection of nonsecreting myeloma cell lines with appropriate expression vectors containing these recombinant genes results in the snythesis and secretion of the chimeric antibody.

There are three major reasons why chimeric antibodies are of great interest. First, mouse hybridomas secreting monoclonal antibodies of essentially any desired specificity exist or can be derived. Hence there is available a potential library of V region genes of enormous diversity. Second, the therapeutic administration of mouse monoclonal antibodies to humans or animals of veterinary interest raises problems of host hypersensitivity. Such reactions may be minimized if the C regions of the administered antibodies are host homologous. Also, most antibody-mediated defense mechanisms invoke such host-based effector mechanisms as transport, the complement system and phagocytosis. All of these involve the constant regions of the immunoglobulin molecule and are likely to operate more efficiently with homologous than with heterologous species constant regions. Third, chimeric molecules in which the antigen binding portion of an immunoglobulin is joined to or incorporates some other functional polypeptide sequence, could find a variety of uses in diagnosis and therapy. For example, appropriate antibody binding sites liked to "reporter" sequences with enzyme activity could be useful in diagnostic and localization procedures. One could also imagine therapeutic applications of enzyme/antibody chimeras. Toxin-mediated therapies directed at tumors or virus infected cells might be precisely targeted by the construction of chimeric molecules in which the binding site for tumor or virus associated determinants is linked to cytotoxic polypeptides.

A number of groups have reported the construction of expression vectors which encode the production of chimeric antibodies and the initial report of Boulianne, Hozumi, and Shulman (1984), announcing the production of TNP-specific antibodies which had mouse V regions and human mu constant regions, was quickly followed by others (Oi et al., 1984). The carefully conceived and executed series of experiments by one group (Bruggemann et al., 1987) has provided a much more precise understanding of the relationship between effector function and Ig class in chimeric mouse-human Igs. These workers constructed and compared a set of chimeric Igs in which the a murine V_H which confers anti-NP specificity was mounted on human IgM, IgG_1 IgG_2, IgG_3, IgG_4, IgE and IgA constant regions. A comparison of the relative efficacies of this matched set of antibodies allowed one to establish that IgM and IgG_1 were more effective than all the other IgG classes in complement-dependent lysis. IgG_1 was also more effective than all other IgG classes in eliciting cell-mediated cytotoxicity.

A novel antigen binding enzyme, in which the capacity to bind the hapten, 4-hydroxy- 3-nitrophenacetyl (NP), is genetically conjugated to the S. aureus nuclease gene has been engineered and expressed by Neuberger, Williams and Fox (1984). This anti-NP immunase retained 10% of the activity of the native S. aureus nuclease and the authors have demonstrated its use in an ELISA assay.

Anti-idiotype Manipulation

Idiotypes are antigenic determinants of variable regions of the antigen receptors of B (immunoglobulin) and T (T cell receptor) lymphocytes. The autoimmunogenicity of idiotypes distinguishes their functional interactions with their host immune system from isotypes and allotypes. Jerne (1974) proposed a network theory of the immune system which focused attention on the immunoregulatory role of idiotype/anti-idiotype interactions. Jerne's original formulation and later modifications and enlargements of the theory recognize that a key feature of the network is the fact that the binding site of an antibody (AB 1) which is complementary to an antigen (AG 1) will be recognized by an anti-idiotype antibody (AB 2). In some instances the anti-idiotype, AB 2, will bear epitopes that are conformationally quite similar to the epitopes of the original antigen, AG 1. A number of interesting demonstrations of the capacity of such antigen mimicking AB 2 preparations to elicit some of the biological effects of the antigen, AG 1, have been described (see Table 1).

Table I. Selected examples of anti-idiotypes as alternatives to antigen*.

Mimickry	Serge and Peterson (1978) reported the preparation of anti-idiotype (anti-anti-insulin and anti-anti-retinol binding protein) that elicited biological effects that were characteristic of the antigens used to raise the immunizing idiotypes. This report led to the concept of "homobodies", anti-idiotypes that bear eptiopes sufficiently similar to those of the original antigen that administration of the anti-id mimics the effects of the antigen.
Parasites	Sacks and Sher (1983) made the first demonstration that anti-id induction of a humoral immune response can substitute for antigen in the induction of protective immunity. They showed that the injection of mice with anit-id raised against three protective monoclonal idiotypes reactive with *Trypanosoma rhodesiense* induced complete or partial immunity to infection with this parasite.
Viruses	Kennedy et al., (1983) have shown that a polyclonal rabbit anti-id raised against human antibody to type B viral hepatitis antigen (HB) protected chimpanzees against HBV challenge. The ability of anti-idiotype to induce cell-mediated immunity in mice to Reovirus and Sendai virus was shown by Fields, Greene and their associates (Sharpe et al., 1984 who used monoclonal, anti-id to elicit a reovirus hemagglutinin-specific DTH and anti-idiotype bearing hybridoma cells to generate CTL activity specific for reovirus infected cells.
Bacteria	McNamara et al., (1984) showed that a monoclonal anti-id induced a protective anti-*S. pneumoniae* response in mice.

(*This list is representative but does not include all known reports.)

In 1981 (Nisonoff and Lamoyi and, independently, Roitt et al.) it was proposed that internal image antibodies might be exploitable as a new type of vaccine. The idea was to make first an antibody (AB 1) to antigenic material from the pathogen in question and then make a second anti-idiotype antibody (AB 2) to the idiotype of the first. They went on to point out that if AB 2 was antigen mimicking, it might be used *in lieu of* the pathogen or its antigens of the formulation of a vaccine. Since then work has progressed rapidly toward using anti-id based vaccines as alternatives to conventional forms of vaccination against parasites, viruses and bacteria. Such an approach is attractive, particularly when combined with hybridoma technology, for the following reasons:

- In cases where it is difficult or impractical to obtain adequate amounts of the pathogen or its antigenic determinants, antigen-mimicking anti-idiotypes can be produced in large amounts by monoclonal antibody technology.

- Anti-idiotype strategies are the only routes available to the biotechnologist when the critical antigenic determinants to the pathogen are carbohydrate or lipoidal and hence inaccessible by peptide synthesis or recombinant DNA techniques.

- The current regulatory climate may be more accepting of monoclonal antibodies than of vaccines employing live recombinant organsism.

What to Expect

A review of these newer developments shows a clear evolutionary trend in the technology of antibody production. Serology began and remained for many years an art, albeit scientifically practiced, in which one performed certain procedures involving antigens, animals and adjuvants and obtained all too finite quantities of antigen-reactive serums. In skilled hands these sera could be absorbed and standardized to produce still ill-defined, but extremely useful, polyclonal antisera of great analytical power. When one stripped away all the ingenious absorption protocols and elaborate immunochemical procedures of concentration and affinity purification, it was clear that investigators took what the aminals gave and did the best they could. With the invention of monoclonal antibody technology by Kohler and Milstein, a dramatic change began. For the first time, the investigator could introduce an important element of control over the definition of the antibodies obtained. Instead of a mixture of many antibodies, a single immunoglobulin of defined specificity was available without any program of absorption whatsoever. Furthermore, once a hybridoma producing a useful monoclonal antibody had been derived, its monoclonal antibody product could be obtained in unlimited amounts for the indefinite future. No longer was it required for some acceptable version of the wheel to be reinvented again and again. But as we have seen, the watershed year of 1975 was only the beginning, a point of departure.

Since its introduction, hybridoma technology has undergone a series of modifications, all of which provide the experimenter with ever greater control and an increasing freedom from the biological constraints inherent in conventional immunization procedures. The powerful methods of partner selection provided by receptor-mediated focusing of B-cells as well as the alternative selection of hybrids of desired specificity by FACS makes the construction and selection of hybridomas much less a lottery. *In vitro* immunization frees one from animal immunization and many of the constraints imposed by tolerance while at the same time, dramatically, lowering the requirements for antigen. The demonstrations of the feasibility of chimeric antibody construction have made it clear that antibody combing sites are for hire. Thus V regions and C regions from diverse sources may be genetically conjugated to each other or even to other functional proteins to produce hybrid molecules yet unrealized by evolution.

One must expect that things will go farther still. Increasingly, one will see the wider application of *in vitro* mutagenesis (Zoller and Smith, 1983) to the deliberate tailoring of antibody combining sites. A fusion of hybridoma methodology and recombinant DNA technology has already begun to transform the antibody from the product of a specialized cell into a creation of genetic engineer. Such developments during this golden age of immunology presage the death of serology as we knew it. They mark the coming of an era of rationally designed molecules which are engineered to bind particular conformations with just the desired binding affinity. But in many cases, the binding will be the beginning not the end. Other functional capabilities, enzymatic, cytotoxic or regulatory, will be grafted to binding sites to produce families of precisely targetable and versatile polypeptides.

References

Andersson, J. and Melchers, F. 1978. The antibody repertoire of hybrid cell lines obtained by fusion of X63-AG8 myeloma cells with mitogen activated B-cell blasts. Contempory Topics in Microbiology and Immunology 81: 1520.

Bazin, H. 1982. Production of rat monoclonal antibodies with the LOU rat nonsecreting IR983F myeloma cell line. In *Protides of the*

Biological Fluids. Ed. H. Peters, 29th Colloquium p 615. Pergamon Press, Oxford and New York.

Boulianne, G.L., Hozumi, N. and Shulman, M.J. 1984. Production of functional chimaeric mouse/human antibody. Nature 312: 643.

Ertl, H.C.J. and Finberg, R.W. 1984. Sendai virus-specific T-cell clones: induction of cytolytic T. cells by anti-idiotypic antibody directed against a helper T-cell clone. Proc. Natl. Acad. Sci. USA 81: 2850.

Galfre, G., Milstein, C. and Wright, B. 1979. Rat X rat hybrid myelomas and monoclonal anti-Fd portion of mouse IgG. Nature 277: 131.

Goldsby, R.A., Osborne, B.A., Suri, D., Mandel, A., Williams, J., Gronowicz, E. and Herzenberg, L.A. 1978. Production of specific antibody without specific immunization. Contemporary Topics in Microbiology and Immunology, 81: 149.

Goldsby, R.A. 1986. Newer developments in hybridoma technology. In: Foodborne Microorganisms and Their Toxins: Developing Methodology. Eds. N. Stern and M. Pierson. p. 241. Marcel Dekker, New York.

Jencks, W.P. 1969. "Catalysis in Chemistry and Enzymology", McGraw Hill, New York, p 288.

Jerne, N.K. 1974. Toward a network theory of the immune system. Ann. Immunol. (Paris) 125C: 373.

Kennedy, R.C., Alder-Storthz, K., Henkel, R.D., Sanchez, Y., Melnick and J.L., Dreesman, G.R. 1983. Immune response to hepatitis B surface antigen: Enhancement by prior injection of antibodies to the idiotype. Science 221: 853.

Klinman, N.P. and Linton, P.J. 1988. The clonotype repertoire of B cell subpopulations. In advances in Immunology, Vol. 42. p 1.

Kohler, G. and Milstein, C. 1975. Continuous cultures of fused cells secreting antibody of predefined specificity. Nature 256: 495.

Lo, M.M.S., Tsong, T.Y., Conrad, M.K., Strittmatter, S.M., Hester, L.D. and Synder, S.H. 1984. Monoclonal antibody production by receptor-mediated electrically induced cell fusion. Nature 310: 792.

Luben, R.A. and Mohler, M.A. 1982. *In vitro* immunization as an adjunct to the production of hybridomas producing antibodies against the lymphokine osteoclast activating factor. Mol. Immunol. 17: 635.

McHugh, Y. 1984. *In vitro* immunization for hybridoma production. In: Hybridoma Technology in Agricultural and Veterinary Research, N.J. Stern and H.R. Gamble (eds.) Rowman and Allanheld Totowa, New Jersey, p. 216.

McNamara, M.K., Ward, R.E. and Kohler, H. 1984. Monoclonal idiotope vaccine against *Streptococcus pneumoniae* infection. Science 226: 1325.

Morrison, S.L., Johnson, J.M. and Herzenberg, L.A., Oi, V.T. 1984. Chimeric human antibody molecules: Mouse antigen-binding domains with human constant region domains. Proc. Natl. Acad. Sci. 81: 6851.

Napper, A.D., Benkovic, S.J., Tramontano, A. and Lerner, R.A. 1987. A stereospecific cyclization catalyzed by an antibody. Science 237: 1041.

Neuberger, M.S., Williams, G.T. and Fox, R.O. 1984. Recombinant antibodies possessing novel effector functions. Nature 312: 604.

Nisonoff, A. and Lamoyi, E. 1981. Hypothesis: Implications of the presence of an internal image of the antigen in anti-idiotypic antibodies: Possible application to vaccine production. Clin. Immun. and Immunop. 21: 397.

Olsson, L. and Kaplan, H.A. 1980. Human-human hybridomas producing monoclonal antibodies of predefined specificity. PNAS 77: 5429.

Parks, D.R., Bryan, V., Oi, V.T. and Herzenberg, L.A. 1979. Antigen-specific identification and cloning of hybridomas with a flourescence-activated cell sorter. Proc. Natl. Acad. Sci. 76: 1962.

Pollack, S.J., Jacobs, J.W. and Schultz, P.G. 1986. Selective chemical catalysis by an antibody. Science 234: 1570.

Raybould T.J., Wilson, P.J., McDougall, L.J. and Watts, T.C. 1985. A porcine-murine hybridoma that secretes porcine monoclonal antibody of defined specificity. Am. J. Vet. Res. 46: 1768.

Raybould, T.J. and Takahashi, J. 1988. Production of stable rabbit-mouse hybridomas that secrete rabit mAB of defined specificity. Science 240: 1788.

Sacks, D.L. and Sher, A. 1983. Evidence that anti-idiotype induced immunity to experimental African Trypanosomiasis is genetically restricted and requires recognition of combining site-related idiotypes. J. Immunol. 131: 1511.

Schultz, P.G. 1988. The interplay between chemistry and biology in the design of enzymatic catalysis. Science 240: 426.

Sege, K. and Peterson, P.A. 1978. Use of anti-idiotypic antibodies as cell-surface receptor probes. Proc. Natl. Acad. Sci. USA 75: 2443.

Sharpe, A.H., Gaulton, G.N., McDade, K.K., Fields, B.N. and Greene, M.I. 1984. Syngeneic monoclonal anti-idiotype can induce cellular immunity to reovirus. J. Exp. Med. 160: 1195.
Sigal, N.H. and Klinman, N.R. 1978. The B-Cell clonotype repertoire. Annual Reviews of Immunology 26: 255.
Sugasaware, R.J. 1988. Hybridoma growth factors. Biotechnology 6: 895.
Thomson, K.M. 1988. Human monoclonal antibodies. Immunology Today 9: 113.
Tramontano, A., Janda, K.D. and Lerner, R.A. 1986. Catalytic antibodies. Science 234: 1566.
Trenkner, E. and Riblet, R. 1975. Induction of Antiphosphorylcholine antibody formation by anti-idiotypic antiobodies. Jour. of Exp. Med. 142: 1121.
Zimmermann, U. and Vienken, J. 1984. Electrofusion of cells. In: Hybridoma Technology in Agricultural and Veterinary Research, N.J. Stern and H.R. Gamble (eds.) Rowman and Allanheld Totowa, New Jersey, p. 173.
Zoller, M.J. and Smith, M. 1983. Oligonucleotide directed mutagenesis of DNA fragments cloned into M13 vectors. In: Methods of Enzymology vol 100, R. Wu, L. Grossman and K. Moldave (eds.), Academic Press, New York, p. 468.

Summary

The subject of this session, Evaluation of Food Quality, is at once complex and challenging. Quality parameters routinely evaluated range from moisture content thru nurtitional components and processability to considerations of safety as impacted by chemical and microbial contaminants. A diverse set of established and emerging technologies are being brought to bear on the challenges associated with food quality determinations as evidenced by the presentations in this session.

Microbial contamination concerns, routinely monitored by classical microbiological techniques, are now being monitored at the molecular level through the use of nucleic acid probes that identify the organism of interest. As described in this session, a series of sophisticated, yet relatively simple, rapid analyses have been developed which permit the specific detection of potential pathogens in food products. The use of nucleic acid-specific probes in these analyses allows the detection of low levels of contaminants with a specificity at least equal to that achieved using standard culture methods. As additional probes are developed this technology should permit an even broader range of microbial evaluations to be conducted extending beyond the bacteria and yeasts to include viruses which are poorly monitored today.

The improvement of food quality through directed breeding of species with improved traits is being enhanced through the development of detailed geonomic maps with readily identifiable markers termed Restriction Fragment Length Polymorphisms (RFLPs). Combined with the techniques of genetic engineering, this new technology can help the breeder create new genetic variability as well as more efficiently utilize existing genetic variations to improve desired traits. As discussed here, RFLP technology should permit plant breeders to develop improved crop plants in shorter time frames while targeting trait improvements more specifically. This becomes even more critical where breeders deal with quantitative (multi-gene) genetic traits. Another potential application of RFLP technology is in the "fingerprinting" of species at the nuclei acid level to confirm the integrity of selected germplasm to assure high quality seed for the grower. This application of RFLP technology is being used in mammalian systems for the diagnosis of genetic disorders and some legal applications.

Another technology, one which has been in routine use for a number of years but is now experiencing a resurgence of development and applicability, is hybridoma technology. Recent refinements of this technology have made possible *in vitro* stimulation of cells to produce useful antibodies. This means that very small amounts of rare antigens, and even antigens potentially toxic to whole animals, can be used to develop monoclonal antibodies under controlled cell culture conditions. The use of some of the recently identified cytokines permits regulation of the cultured cells. Another development that can lead to specific monoclonal antibody production is receptor mediated cell "partner formation." Coupled with improved methods to initiate hybridoma formation through cell fusion, such as electrofusion, this technology will continue to evolve and find new applications. An exciting potential extension of this technology is the creation of chimeric antibodies. These chimeras combine the specificity of the antibody binding site with enzyme activity within the immunoglobulin molecule. These "abzymes" hold promise as a potent new tool of biotechnology.

The presentations in this session provide excellent insights into established and emerging technologies with application to the problems of evaluating food quality. The future of this area, as evidenced by the content of this session, indeed appears bright.

Richard Holsten
Manager of Biotechnology
E.I. du Pont de Nemours & Co., Inc.
Wilmington, DE, USA

BIOPROCESSING

Bioprocessing of Meats

L. Leistner and F.-K. Lücke
Institute for Microbiology, Toxicology and Histology
Federal Center for Meat Research
8650 Kulmbach, Federal Republic of Germany

Fermented meat products have been known for centuries, and are increasingly liked today, since they are "naturally" preserved. All meats for which microorganisms (bacteria, yeasts or molds) could be benefical for their preservation, flavor or appearance should be thoroughly studied, with the intention to optimize their "bio-preservation," if possible by decreasing other additives. In using better control devices for the extrinsic factors (temperature, relative humidity, air velocity, etc.) and intrinsic factors (a_w, pH, pO_2, etc.) of fermentation, an optimization and the automation of processes becomes feasible. During fermentation microorganisms in the interior of meats are immobilized in the matrix, a fact which must be taken into account when investigating growth kinetics and metabolism of, and interactions between, microorganisms during sausage fermentation. The use of starter cultures contributes to the control of various meat fermentations, and the hitherto available strains could be improved by better selection, as well as mutation and genetic engineering.

Introduction

Fermented sausages are raw meat products stabilized by a reduced a_w and pH. In Europe raw sausages are fermented with microorganisms, mainly lactic acid bacteria and apathogenic staphylococci. This process has been known for about 250 years and probably originated in Italy (Leistner, 1986a). The Chinese raw sausage (Lup Cheong) has been known for more than 1,500 years. This product is stabilized by a_w only, because a low pH caused by fermentation would be undesirable (Leistner, 1988). Raw hams, which are known in Europe and China for at least 2,500 years (Leistner, 1986b), are stable due to a reduced a_w and generally are not fermented. Since the bacterial count in the interior of high quality raw hams is low, bacteria contribute little to the flavor and stability of such products. However, if raw hams are cured without refrigeration (e.g., for Turkish Pastirma dried beef), then lactic acid bacteria may contribute to the safety of the meat (El-Khateib et al., 1987). It also has been suggested to stabilize raw hams made from DFD-meat by the injection of lactobacilli which grow at 8°C (Hammes and Arnold, 1986). Some fermented sausages (e.g. Italian or Hungarian salami) and raw hams (e.g. Bündnerfleisch of Switzerland, Coppa as well as Südtiroler Bauernspeck of Italy) are mold-fermented. The desirable mold growth on the surface of such meats improves the appearance, flavor and preservation of these products (Leistner, 1986c).

The flavor of raw and cooked hams may be influenced by bacteria present in the cover brine (Leistner, 1958; Peräjä et al., 1973) or by bacteria injected into the hams. Also for Bologna-type sausage (Brühwurst) the use of starter culture bacteria has been suggested (Petäjä, 1977; Schiefer and Schöne, 1980) with the intention to improve the flavor and the color of the products. For Gelderse Rookworst (Brüthwurst), a product of the Netherlands which formerly was stabilized by lactic acid bacteria, now the addition of GdL (glucono-delta-lactone) is preferred (Leistner, 1985a, 1987). It was also considered that the addition of selected "protective" cultures to fresh meat to be vacuum-packaged (Hanna et al., 1980; Schillinger and Lücke, 1986; Renerre and Montel, 1986) and to minced meat (Reddy et al., 1970; Fetlinski et al., 1979; Gibbs, 1987) may contribute to the control of undesirable bacteria.

In general, all processes for the preservation of meat in which desirable microorganisms occur should be thoroughly investigated, with the intention to optimize the appearance, flavor and shelflife of the products by using suitable cultures and at the same time to reduce the addition of substances, such as nitrate/nitrite or sodium chloride, which are less desirable from the toxicological or nutritional point of view. Such "bio-preservation" of meats would find general approval, especially of the younger generation of consumers in Europe.

Technology of Fermentation

The traditional fermentation processes for meats have been developed over centuries by trial and error. The process is designed to give desirable microorganisms an advantage and suppress undesirable microorganisms responsible for spoilage and food-poisoning.

The extrinsic factors most important for proper fermentation of meats are the temperature, relative humidity and air velocity, as well as the time these parameters are applied to the product. Of the intrinsic factors the a_w, pH and pO_2 (partial pressure of oxygen) are of paramount importance. Much is known about the a_w, and PH of fermented sausages and raw hams. However, it is quite difficult to measure O_2 in solid substrates. The extrinsic and intrinsic factors important for the fermentation of raw sausages have been recently reviewed by Rödel (1985). As Leistner (1986 a,b; 1987) has pointed out, the stability of raw sausages and hams is due to a sequence of barriers active in the products. For example, in fermented sausages the hurdles are nitrite curing salt, followed by pO_2, competitive organisms, pH and a_w, whereas in raw hams the pH, temperature and a_w secure stability. Furthermore, in accordance with the HACCP-concept, guidelines have been suggested for the production of fermented sausages and raw hams; i.e. 19 critical control points for fermented sausages and 15 for raw hams (Leistner, 1985b).

Over the years the construction and control of the ripening rooms for fermented sausages have become more sophisticated, but at the same time more energy is consumed. Stiebing et al, (1982) suggested a simplification in the essential control of the relative humidity (RH) during the ripening of fermented sausages by using fresh air for RH control. In this manner, as much as 70% of the energy required for the production of fermented sausages was saved. However, even further improvements are feasible. For instance, by the continuous measurement of the pH and a_w inside the product (i.e. in one representative sausage) and by using these data to control the extrinsic factors (climate of the ripening room), the production of fermented sausages could be optimized and automated.

Topography of Fermentation

The extrinsic and intrinsic factors of the ripening process as well as the counts and types of microorganisms important for fermented sausages have been intensively studied. However, it is important to note that the natural flora as well as the added starter cultures are not evenly distributed in a fermented sausage, but are immobilized in cavities of the sausage mix. The distance between these cavities or nests of ripening flora varies between 100 and 5000 µm (Katsarus and Leistner, 1988).

If the properties of a sausage are changing during the ripening process in the desired direction (nitrate reduction, lactic acid production, catalase activity, etc.), then "large areas" of the sausage which are located between these cavities must be influenced by the bacteria growing in the nests. Since the microorganisms are trapped and cannot be released from these nests, the ripening of sausages can be regarded as a solid-state-fermentation (Katsaras and Leistner, 1987).

It must be assumed that the microorganisms which grow in such nests are in keen competition. If such a nest is made up by chance of a pure culture, e.g. of lactobocilli, then the individual bacteria will compete for the nutrients and impair each other with their metabolic products, such as lactic acid. Therefore, after some time the growth ceases, because the cell division will be delayed. We observed this in nests of lactobacilli as well as of apathogenic staphylococci (Katsaras and Leistner, 1987, 1988). Furthermore, inhibitory substances such as lactic acid which are produced in "lactobacilli nests" will diffuse and thus inactivate microorganisms, e.g. salmonellae or pathogenic staphylococci, present in other nests. Whether metabolites such as lactic acid or bacteriocins, will diffuse equally well through the sausage matrix needs further study. Sometimes different types of bacteria are trapped in one cavity and then the competition will be fierce, but lactic acid bacteria will have an advantage due to their tolerance of low pO_2, low pH and low a_w.

An investigation of the topography of fermented meats, using scanning electron microscopy, should lead to a better understanding of the process and, subsequently, their improvement. A relatively small distance between the nests of desirable bacteria in the sausage matrix should be advantageous. Therefore, a more even inoculation of the sausage mix with equal starter cultures might prove more important than previously envisioned.

Microbiology of Fermentation

The microbiology of fermented sausages has recently been reviewed by Lücke (1985). Furthermore, Lücke (1986) has described the microbiological events occurring during the ripening of raw sausages and hams, and Hechelmann (1985) reported on spoilage problems of these products. The significance of food-poisoning organisms for raw sausages and hams was discussed for salmonellae (Schmidt, 1985), *Staphylococcus aureus* (Hechelmann et al., 1988) and *Clostridium botulinum* (Lücke et al., 1982, 1983) as well as for toxigenic molds (Leistner, 1984). Moreover, Lücke and Hechelmann (1987) described the composition and effects of starter cultures recommended for raw sausages and hams.

Excellent fermented sausages can be manufactured without addition of starter cultures. If suitable ripening conditions are maintained, then the desired ripening flora will prevail, even if only a few lactobacilli and micrococci are present in the raw material. Nevertheless, the use of selected starters, added as pure or mixed cultures, is often beneficial to the quality consistency and stability of fermented sausages.

In the United States, the importance of lactobacilli for the fermentation of raw sausage was recognized early (Jensen and Paddock, 1940). However, since lactobacilli proved initially difficult to lyophilize, *Pediococcus cerevisiae* (reclassified as *P. acidilactici*) was introduced as starter culture (Deibel et al., 1961a,b). Today many starter cultures for fermented sausage contain pediococci (*P. pentosaceus, P. acidilactici*). Starter preparations may also contain lactobacilli; especially *Lactobacilus plantarum*, but also *L. sake* and *L. curvatus*. The main function of lactic acid bacteria in the fermentation of meat is the production of lactic acid, which lowers the pH and stabilizes the food. However, many strains of lactic acid bacteria may form additional bacteriostatic compounds. These include peroxides (in the presence of oxygen), acetic acid (from gluconate, pentoses or, if oxygen is present, from hexoses) and bacteriocin-like compounds (Schillinger and Lüke, 1988).

In Europe, a suitable starter culture for meat products was first introduced by Niinivaara (1955). *Micrococcus* strain (M 53) proved beneficial because it rapidly reduced nitrate, improved the color and flavor and inhibited undesirable bacteria (Pohja and Niinivaara, 1957). This strain was later replaced by a "fermentative *Micrococcus*" isolated by Pohja (Licentiate Thesis, Helsinki, 1960) and other *Micrococcaceae*, including *Staphylococcus carnosus* (renamed by Schieifer and Fischer, 1982). *Micrococcaceae* are beneficial due to the production of nitrate reductase and catalase. Catalase formation destroys peroxides and prevents color and flavor defects in fermented meats. The lipolytic properties of some *Micrococcaceae* are apparently important for the flavor development in long-ripened fermented sausage, such as genuine Italian salami. Nurmi (1966) recommended a combination of *L. plantarum* and Pohja's "fermentative *Micrococcus*," because this controlled mixed culture caused a rapid reduction in pH without inhibiting color and flavor development. Combination of lactic acid bacteria and *Micrococcaceae* are now widely used as starters for fermented sausages.

Yeasts occur frequently in fermented sausages, especially on the surface of unsmoked products. *Debaryomyces* spp. are predominantly found (Leistner and Bem, 1970; Comi and Cantoni, 1980). *Candida spp.* are also found, but several species of the latter genus are potentially pathogenic. Yeasts have been observed to accelerate the color formation in sausages and may improve their flavor (Miteva et al., 1986) and appearance. In France, some yeast growth ("fleur de saucisson") is regarded as desirable and, therefore, yeast starter cultures for surface inoculation are commercially available. Rossmanith et al., (1972) observed that color and flavor formation in fermented sausages could be improved by addition of selected *Debaryomyces* strains to the sausage mix. Coretti (1977) recommended for this purpose a combination of *D. hansenii*, lactobacilli and micrococci as starter cultures.

Mold contribute to the characteristic aroma, flavor and appearance of mold-ripened sausages and hams. The mold growth on the surface might also delay the rancidity of the products. Traditionally, mold colonization of mold-fermented meats was achieved by the "house-flora" of the ripening rooms. Molds found on such products are predominantly penicillia (Leistner, 1986c). However, since most penicillia are toxigenic (Leistner, 1984), starter culture for mold-fermented meats was developed by Mintzlaff and Leistner (1972). Today in Germany and France, six such starter cultures are on the market representing three biotypes of *Penicillium nalgiovense*. These starter cultures should not only prevent mycotoxin formation in meat products, but also suppress undesirable molds responsible for color and flavor defects in mold-fermented meats.

Protective Cultures

A general distinction could be made between "starter cultures" which improve the sensory properties of a food, and "protective cultures," which suppress undesirable microorganisms such as salmonellae, pathogenic staphylococci, *Listeria monocytogenes*, *Clostridium botulinum*, and toxigenic molds. However, both purposes could be fulfilled by one culture or a controlled mixed culture.

Undoubtedly, for suppression of salmonellae in fermented sausages, especially in quick ripened products, starter cultures containing lactobocilli are very useful. Schmidt (1987) achieved the best control of salmonellae if lactobacilli and GdL were added simultaneously.

Lactobacilli probably form acetic acid from gluconate, thus stimulating the inhibitory effects. Schillinger and Lücke (1988) confirmed that a rapid initial rate of acid formation is essential to control salmonellae during sausage fermentation. No evidence was found of the involvement of other inhibitory metabolic products of lactic acid bacteria.

If fermented sausages are ripened at elevated temperatures, then *Staphylococcus aureus* and its enterotoxin formation are a risk to the consumer. *S. aureus* grows much better in the surface layer than in the core of fermented sausages (Barber and Deibel, 1972), since this organism tolerates the inhibitory factors (pH, a_w, nitrite, etc.) of sausages better at a higher partial pressure of oxygen. The growth of *S. aureus* in fermented sausages is inhibited by chemical acidulation (addition of GdL) and/or by microbial acid formation (Daly et al., 1973). Consequently, the guidelines issued by the American Meat Institute in 1982 specify, for each temperature, the maximum time available for the product to reach pH$\leq$5.3. It must be emphasized that *S. aureus* may grow better on mold-fermented sausages since the surafce flora increases the pH and the antimicrobial compounds associated with smoke are absent. Such products must be ripened at lower temperatures. Hechelmann et al., (1988) have recently published critical control points for the manufacture of unsmoked, mold-fermented sausages.

Clostridium botulinum (types A, B and E) "may be less at a risk" in fermented sausages common in Germany. This was demonstrated by a thorough investigation conducted in our laboratory (Hechelmann et al., 1980; Lücke et al., 1983). Multiplication and toxin formation of *C. botulinum* is inhibited by the combined effect of pH, a_w and the antagonistic behavior of the usual ripening bacteria. Nitrite was not necessary for this inhibition. The formation of botulinum toxin was reported by Christiansen et al, (1975) in "summer sausage" of the U.S. if ripening temperatures above 30°C were employed and no sugar was added; the pH remained high. Also Incze and Delényi (1979) observed botulinum toxin formation in Hungarian salami, but only after injection of spores into the sausages. It deserves further investigation to determine whether reduction of sodium chloride in fermented sausages increase the risk of botulism. If it does, increasing the rate and extent of pH drop, use of alternative a_w-lowering agents (such as freeze-dried meat) and/or lowering the ripening temperature should be considered as alternatives to eliminate any microbial hazard.

New Starter Cultures

In the case of mold-fermented meats, the possible formation of mycotoxins by mold starter cultures hitherto regarded as safe is an important field of study (Leistner, et al., 1988). After a reliable toxicological protocol to evaluate mold starter cultures was established (Fink-Gremmels et al, 1988), the commercially available starter cultures are now under investigation for mycotoxin formation, including those mycotoxins for which analytical standards are not yet available. Our laboratory has developed two effective and safe cultures for mold-fermented meats, both representing *P. nalgiovense*.

New bacterial starter cultures are also being developed to improve both the sensory quality and the microbiological safety of fermented meats. For instance, by using bacterial starter cultures that are strongly antagonistic against spoilage bacterial and pathogens, it may become feasible to produce low-acid, low-salt fermented meats that are less susceptible to microbial hazards.

The first step in the development of better starter cultures for fermented meats is the precise definition of those properties a starter strain should have and not have. Then a large number of organisms can be screened for these properties. This "classical" approach is by no means made obsolete by the introduction of modern genetic techniques. If the desired features are not found in organisms from appropriate habitats, strain improvement by genetic methods could prove advantagous. A prerequisite for the use of genetic engineering techniques for strain improvement is a thorough knowledge of the biochemical basis of the desired and undesired characters. For many of such properties (e.g. growth at low temperatures), this is very difficult to elucidate; hence, classical (undirected) mutation and selection techniques will continue to be employed.

In the last few years, genetic studies have been initiated on virtually all bacterial species available as starter cultures for meat fermentations. These organisms include the lactobacilli and pediococci (see Chassy, 1987, and Sandine, 1987, for recent reviews), *Staphylococcus carnosus* (Götz, 1986) as well as *Penicillium nalgiovense*. In our laboratory, Geisen et al. (1988) recently developed a transformation system for *P. nalgiovense* which offers promise for use as a recipient for isolated genes and the improvement of this mold via genetic methods (Leistner et al., 1988).

Psychrotrophic lactobacilli (*L. sake, L. curvatus*) dominate on chilled meats and in the microflora of sausages fermented at lower temperatures. Therefore, selected strains of these species may be used as "protective cultures" to suppress undesired microorganisms. Romero and McKay (1985) have detected a plasmid in one of these strains and demonstrated conugal transfer of the pAMß plasmid. In an investigation on 37 *L. sake* strains, we found a large diversity both of phenotypes and plasmid content (Lücke et al., 1987) and obtained evidence that certain plamids carry genes that may be useful as markers for construction of "food-grade" vectors (Lücke et al., 1988).

Knowledge of the molecular biology of microbial species involved in meat fermentation could lead to a new generation of starter cultures and possibly to a wider range of fermented and/or biologically preserved meats. Recombinant DNA techniques are a powerful tool to gain such knowledge. However, if too much emphasis is given to the "construction" of organisms by recombinant DNA techniques, the public acceptance of meats processed with novel cultures could be seriously impaired. To avoid this, "natural" transfer of genes (i.e., by conjugation and parasexual recombination) should be attempted first. It will be wise to avoid any use of genes from non-food-grade microorganisms (such as *E. coli*) and antibiotic resistance markers in the "construction" of new starter cultures.

References

Barber, L.E. and Deibel, R.H. 1972. Effect of pH and oxygen tension on staphylococcal growth and enterotoxin formation in fermented sausage. Appl. Microbiol. 24: 891.

Chassy, B.M. 1987. Prospects for the genetic manipulation of lactobacilli. FEMS Microbiol. Reviews 46: 297.

Christiansen, L.N., Tompkin, R.B., Shaparis, A.B., Johnston, R.M and Kautter, D.A. 1975. Effect of sodium nitrite and nitrate on *Clostridium botulinum* growth and toxin production in a summerstyle sausage. J. Food. Sci. 40: 488.

Comi, G. and Cantoni, C. 1980. Yeasts in matured raw sausages. Industrie Alimentari 19: 857 (in Italian).

Coretti, K. 1977. Starterkulturen in der Fleischwirtschaft. Fleischwirtschaft 57: 386.

Daly, C., La Chance, M., Sandine, W.E. and Elliker, P.R. 1973. Control of *Staphylococcus aureus* in sausage by starter cultures and chemical acidulation. J. Food Sci. 38: 426.

Deibel, R.H., Niven, C.F. Jr. and Wilson, G.D. 1961a. Microbiology of meat curing. III. Some microbiological and related technological aspects in the manufacture of fermented sausages. Appl. Microbiol. 9: 156.

Deibel, R.H., Wilson, G.D. and Niven, C.F. Jr. 1961b. Microbiology of meat curing. IV. A lyophilized *Pediococcus cerevisiae* starter culture for fermented sausages. Appl. Microiol. 9: 239.

El-Khateib, T., Schmidt, U. and Leistner, L. 1987. Mikrobiologische Stabilitat van turkischer Pastirma. Fleischwirtschaft 67: 101.

Fetlinski, A., Knaut, T. and Kornacki, T. 1979. Einsatz von Milchsurebokterien als Startukulturen zur Haltbarkeitsverlangerung von Hackfleisch. Fleischwirtschaft 59: 1729.

Fink-Germmels, J., El-Banna, A.A. and Leistner, L. 1988. Entwicklung von Schimmelpilz-Starterkulturen für Fleischerzeugnisse. Fleischwirtschaft 68: 24.

Geisen, R., Glenn, E. and Leistner, L. 1988. Entwicklung eines Transformationssystems für *Penicillium nalgiovense*. Poster P-210, presented at the Srping Meeting of VAAM and DGHM Section I, March 20 - 23, at Regensburg, West Germany.

Gibbs, P.A. 1987. Novel uses for lactic acid fermentation in food preservation. J. Appl. Bact. Sympos. Suppl. 51S.

Götz, F. 1986. Ein neues Wirt-Vektor-System bei *Staphylococcus carnosus*. Umschov 10: 530.

Hammes, W.P. and Arnold, S. 1986. Verfahren zum Herstellen von rohgepökeltem Fleisch in Stuckform, DE 3502063 AI, Offenlegungsschrift des Deutschen Patentamtes vom 24. Juli 1986.

Hanna, M.O., Hall, L.C., Smith, G.C. and Vanderzant, C. 1980. Inoculation of beef steaks with lactobacillus species before vacuum-packaging. I. Microbiological considerations. J. Food Protec. 43: 837.

Hechelmann, H., Lücke, F.-K. und Leistner, L. 1980. Bedeutung von Clostridium botulinum für Rohwurst und Rohschinken. Proceedings 1st Wrold Congress on Foodborne Infections and Intoxications, held June 29 - July 3, 1980 at Berlin (West), p. 823.

Hechelmann, H. 1985. Mikrobiell verursachte Fehlfabrikate bei Rohwurst und Rohschinken. In "Mikrobiologie und Qualität von

Rohwurst and Rohschinken." S. 103. Herausgegeben vom Institut für Mikrobiologie, Toxikologie und Histologie der Bundesanstalt für Fleischforschung, Kulmboch.

Hechelmann, H., Lücke, F.-K. und Schillinger, U. 1988. Ursachen und Vermeidung von Staphylococcus oureus-Intoxikationen nach Verzehr von Rohwurst und Rohschinken. Mitteilungsblatt Bundesanstalt für Fleischforschung, No. 100: 7956.

Incze, K. and Delényl, M. 1979. Influence of additives and ripening parameters on growth and toxin production of Clostridium botulinum. Proceedings 25th E.M.M.R.W., held Aug. 27 - 31, 1979 in Budapest, Hungary, Vol. III, p. 879.

Jensen, L.B. and Paddock, L. 1940. Sausage treatment with lactobacilli. U.S. Patent 2.225.783.

Katsaras, K. und Leistner, L. 1987. "Solid-State-Fermentation" von Rohwurst. Mitteilungsblatt Bundesanstalt für Fleischforschung, No. 96: 7497.

Katsaras, K. und Leistner, L. 1988. Topographie der Bakterien in der Rohwrust. Mitteilungsblatt Bundesanstalt für Fleischforschung, No. 100: 7965.

Leistner, L. 1958. Bakterielle Vorgäng bel der Pökelung von Fleisch. II. Günstige Beeinflussung von Farbe, Aroma und Konservierung durch Mikroorganismen. Fleischwirtschaft 10: 226.

Leistner, L. und Bem, Z. 1970. Varkommen und Bedeutung von Hefen bei Pökelfeischwaren. Fleischwirlschaft 50: 350.

Leistner, L. 1984. Toxigenic penicillia occurring in feeds and foods: a review. Food Technol. Australia 36: 404.

Leistner, L. 1985a. Hurdle Technology applied to meat products of the Shelf Stable Product and Intermediate Moisture Food types. In "Properties of Water in Foods" (D. Simatos and J.L. Multon, eds.). p. 309. Martinus Nijhoff Publisher, Dordrecht.

Leistner, L. 1985b. Empfehlungen für sichere Produkte. In "Mikrobiologie und Qualität von Rohwurst und Rohschinken." S. 219. Herausgegeben vom Institut für Mikrobiologie, Toxikologie und Histologie der Bundesanstalt für Fleischforschung, Kulmbach.

Leistner, L. 1986a. Allgemeines über Rohwrust. Fleischwirtschaft 66: 496.

Leistner, L. 1986b. Allgemeines über Rohschinken. Fleischwirtschaft 66: 496.

Leistner, L. 1986c. mold-ripened foods. Fleischwirtschaft 66: 1385.

Leistner, L. 1987. Shelf-Stable Products and Intermediate Moisture Foods based on meat. In "Water Activity: Theory and Applications to Food" (L.B. Rockland and L.R. Beuchat, eds.). p. 295. Marcel Dekker, Inc., New York and Basel.

Leistner, L. 1988. Shelf-stable Oriental meat products. Proceedings 34th International Congress of Meat Science and Technology, held Aug. 29 - Sept. 2, 1988 at Brisbane, Australia, Part B, p. 470.

Leistner, L., Geisen, R. and Fink-Gremmels, J. 1988. mold-fermented foods of Europe: hazards and developments. Proceedings 7th International IUPAC-Symposium, held Aug. 16 - 19, 1988 at Tokyo, Japan (in press).

Lücke, F.-K., Hechelmann, H. and Leistner, L. 1982. Botulismus nach Verzehr von Rohschinken - Experimentelle Untersuchungen. Fleischwirtschaft 62: 203.

Lücke, F.-K., Hechelmann, H. and Leistner, L. 1983. Fate of Clostridium botulimun in fermented sausages processed with and without nitrite. Proceedings 24th E.M.M.W., held Aug. 28 - Sept. 2, 1983 at salsomoggiore, Italy, p. 403.

Lücke, F.-K. 1985. Fermented sausage. In "Microbiology of Fermented Foods" (B.J.B. Woods, ed.), Vol. 2, p. 41. Elsevier Applied Science Publishers, London.

Lücke, F.-K. 1986. Microbiological processes in the manufacture of dry sausage and raw ham. Fleischwirtschaft 66: 1505.

Lücke, F.-K. and Hechelmann, H. 1987. Starter cultures for dry sausages and raw ham - composition and effect. Fleischwirtschaft 67: 307.

Lücke, F.-K., Berg, C.M., Buckley, L. und Schillinger, U. 1987. Plasmids in *Lactobacillus sake* from meats (Abstract). FEMS Microbiology Reviews 46: P19.

Lücke, F.-K., Berg, C.M., Schillinger, U., Angermann, K. und Buckley, L. 1988. Plasmidgehalt und Variabilität von Merkmalen bei *Lactobacillus sake*-Stämmen aus Fleisch und Fleischerzeugnissen. Poster P-29, presented at the Srping Meeting of the VAAM und DGHM Section I, March 20 -23 at Regensburg, West Germany.

Mintzlaff, H.-J. und Leistner, L. 1972. Untersuchungen zur Selektion eines technologisch geeigneten und toxikologisch unbedenklichen Schimmelpilz-Stammes für die Rohwurst-Herstellung. Zentralblatt für Veterinärmedizin B 19: 291.

Miteva, E., Kirova, E., Gadjeva, D. and Rodeva, M. 1986. Sensory aroma and taste profiles of raw-dried sausages manufactured with a lipolytically active yeast culture. Die Nahrung 30: 829.

Niinivaara, F.P. 1955. Über den Einfluß von Bakterien-Reinkulturen auf dle Reifung und Umrötung der Rohwurst. Ph.D. Thesis, Unviersity of Helsinki; Acta Agralia Fennica 85: 1.

Nurmi, E. 1966. Effect of bacterial Inoculation on characteristics and microbial flora of dry sausage. Ph.D. Thesis, Unversity of Helsinki, Acta Agralia Fennica 108: 1.

Petäjä, E., Laine, J.J. und Niinivaara, F.P. 1973. Einfluß der Pökellakebakterien auf die Eigenschaften gepökelten Fleisches. Fleischwirtschaft 53: 680.

Petäjä, E. 1977. Untersuchungen über die Verwendungsmöglichkeiten von Starterkulturen bei Brühwurst. Fleischwirtschaft 57: 109.

Pohja, M.S. and Niinivaara, F.P. 1957. Über die Reifung der Rohwurst. III. Mitteilung: Über die antagonistische Wirkung eines Mikrokokkenstammes gegen die in Rohwurst vorkommenden Bakterienstämme. Zeitschrift für Lebensmittel-Untersuchung und -forschung 106: 298.

Reddy, S.G., Hendrickson, R.L. and Olson, H.C. 1970. The influence of lactic cultures on ground beef quality. J. Food Sci. 35: 787.

Renerre, M. and Montel, M.C. 1986. Inoculation of steaks with Lactobacillus species and effect on colour and microbial counts Proceedinsgs 32nd E.M.M.R.W., held Aug. 24 - 29, 1986 at Ghent, Belgium, Vol. I, p. 213.

Rödel, W. 1985. Rohwurstreifung - Klima und andere Einflußgroßen. In "Mikrobiologie und Qualität von Rohwurst und Rohschinken." S. 60. Heresgegeben vom Institut für Mikrobiologie, Toxikologie und Histologie der Bundesantalt für Fleischforschung, Kulmbach.

Romero, D.A. and Mackay, L.L. 1985. Isolation and plasmid characterization of a *Lactobacillus* species involved in the manufacture of fermented sausage. J. Food Prot. 48: 1028.

Rossmanith, E., Mintzlaff, H.-J., Streng, B., Christ, W. und Leistner, L. 1972. Hefen als Starterkulturen für Rohwürste. Jahresbericht Bundesanstalt für Fleischforschung 1972: I 47.

Sandine, W.E. 1987. Looking backward and forward at the practical application of genetic research on lactic acid bacteria. FEMS Microbiol. Reviews 46: 205.

Schiefer, G. und Schöre, R. 1980. Untersuchungen zur Anwendung von Starterkulturen bei der Brühwurstherstellung. Fleisch 34: 34.

Schillinger, U. und Lücke, F.-K. 1986. Milchsäurebakterien-Flora auf vakuumverpacktem Fleisch und ihr Einfluß auf die Haltbarkeit. Fleischwirtschaft 66: 1515.

Schillinger, U. and Lücke, F.-K. 1988. Hemmung des Salmonellenwachstums in frischer, streichfähiger Mettwurst ohne Zuckerstoffe. Fleischwirtschaft 68: 1056.

Schiefer, K.H. and Fishcer, U. 1982. Description of a new species of the genus *Staphylococcus: Staphylococcus carnosus*. Int. J. Systemat. Bacteriol. 32: 153.

Schmidt, U. 1985. Salmonellen, Bedeutung bei Rohwurst und Rohschinken. In "Mikrobiologie und Qualität von Rohwurst und Rohschinken." S. 128. Herausgegeben vom Institut für Mikrobiologie, Toxikologie und Histologie der Bundesanstalt für Fleischforschung, Kulmbach.

Schmidt, U. 1987. Hemmung von Salmonellen durch technologische Maßnahmen. Mitteilungsblatt Bundesanstalt für Fleischforschung, No. 96: 7443.

Steibing, A., Rödel, W. und Klettner, P.-G. 1982. Energieeinsparung bei der Rohwurstreifung. Fleischwirtschaft 62: 1383.

Genetic Modification of Enzymes Used in Food Processing

Bhav P. Sharma
Genencor Inc.
San Francisco, CA 94080, USA

Food enzymes constitute about two-thirds of the nonpharmaceutical enzyme market in the United States. The recombinant DNA technology efforts in this area to date have tended to focus on the following approaches: transferring genes from one type of cell into another more suitable for economical production; yield improvement in enzyme production; modification of protein structure to alter enzyme properties. Examples are enzymes in the carbohydrase, protease, and lipase groups. Progress in these areas is reviewed.

In some cases, the next major goal following the technical success of the recent past is to achieve altered eyzymes that are commercially competitive in performance. In others, a more conventional task, and a challenge nonetheless, is to overcome the economic and regulatory hurdles. For the user of food enzymes, success in these endeavors obviously means better performance at a given price. For the producer, it means a specialty chemical status for food enzymes as some of the food enzymes drift towards commodity status.

Introduction

The advent of genetic engineering technology has presented several new horizons in the roles enzymes may play in commercial biotechnology. The significance of biotechnology extends far beyond patent applications. This paper presents an overview of some of the ways recombinant DNA technologies have been utilized in industrial enzyme technology. Some of the hurdles that must be overcome in future commercialization efforts are also outlined.

Food Enzymes

The annual worldwide sales volume of industrial enzymes was estimated at over $400 million in 1985, out of which $260 million or about two-thirds was for enzymes used for food purposes (European

Chemical News, 1986). The United States accounted for about 60% of the food enzymes sales at about $155 million. The two most important uses of food enzymes are in starch conversion and cheese making (Table 1). Starch enzymes add up to about $85 million. Cheese making utilizes about $40 million of enzymes. This leaves about 19% for other uses, such as meat tenderizing, flavor development, juice or wine clarification.

Table 1. 1985 US food enzyme market

Enzymes	Applications	Annual Sales ($-million)	
Carbohydrases			
Alpha amylase	Starch to glucose	15	
Glucoamylase	Starch to glucose	25	
Glucose isomerase	Glucose to fructose	45	
Other	Fruit juice/wine clarification	5	
Subtotal			90
Proteases			
Rennin/chymosin	Cheese making	40	
Others	Meat tenderizing	15	
Subtotal			55
Lipsases			
	Flavor development in dairy products	10	
Subtotal			10
Total			155

Adapted with permission from Genetic Technology News, 1986.
Copyright 1986 Genetic Technology News.

The above data translate to about a 7% growth during 1980-1985 for worldwide use and about 10% for the United States (Scott, 1980). The better growth rate for the U.S. is due to the use of the amylase and isomerase enzymes for the production of High Fructose Corn Syrup

(HFCS) or iso-glucose which has had an impressive success in sugar applications, such as soft drinks.

Genetic Engineering Technologies

The general trend for most enzyme products has tended towards their becoming less and less of a speciality product and more and more of a commodity product. In other words, the general expectation has been for the enzyme prices to come down with time. Market growth for enzymes has generally come from growth of the application market segment itself. In such an environment, the enzyme supplier has an ever present challenge to maintain profit margins in the face of declining prices. The genetic engineering technologies have provided not only a tool to help meet that kind of challenge but also some new possibilities:

- Production of clearly established but previously uneconomical enzymes for known applications
- Production of new/modified enzymes for known applications
- Production of new/modified enzymes for new applications

Recent progress and potential for these loosely defined categories will be reviewed next. Classical mutagenesis and other technologies have made remarkable contributions to help the enzyme suppliers and users, and will no doubt continue to do so in conjunction with the genetic engineering technologies.

Clearly Established Enzymes for Known Applications

Alpha-amylase and chymosin enzymes from certain sources may be placed in this category.

Alpha-amylase
Bacterial and fungal alpha-amylases have been in commercial use for a long time for the hydrolysis of starch into sugars. To help speed up the hydrolysis, alpha-amylase containing starch slurry is heated to temperatures of 105-110°C for short periods (Reichelt, 1983). Heat stability is, therefore, a desired characteristic in the alpha-amylase used. It was discovered that one of the better alpha-amylases was that

obtainable from *Bacillus stearothermophilus* (Tamuri et al., 1981). It showed better thermostability and acid pH stability than the commercially available *Bacillus licheniformis* enzyme. In addition, the *stearothermophilus* amylase was more stable under conditions of low or no calcium. This is an added advantage because of calcium adversely affects glucoamylase and glucose isomerase enzymes which are used in the latter stages of starch processing.

More recently, another attractive alpha-amylase was identified from *Bacillus megaterium* (Marie-Henrietta et al., 1985). This particular amylase had specially useful activities in the saccharification reactions. In combination with glucoamylase, the *B. megaterium* alpha-amylase offered better overall economies because it could allow saccharifying at higher dissolved solids, increased glucose yields, shorter reaction times, and lower concentration of oligomers in the product. It also catalyzed the conversion of pullulan and cyclodextrins.

Commercial use of these amylase enzymes is closer to reality as a result of the efforts at CPC International (CPC, 1984) and its subsidiary, Enzyme Biosystems Ltd (EBS, 1986). CPC and EBS have filed two of the first petitions to the FDA (United States Food and Drug Administration) for GRAS (generally recognized as safe) status for enzymes produced by a recombinant organism. Both petitions demonstrate the utility of genetic engineering technology for food enzymes. The host organism in this work was *Bacillus subtilis* B1-109 (American Type Culture Collection or ATCC 39,701), eventually derived from *B. subtilis* 168 which is a widely used organism in recombinant DNA research. The host organism does not produce an alpha-amylase. In the first case (CPC, 1984), the alpha-amylase gene material from *B. stearothermophilus* (ATCC 39,709) was used to transform the host into the desired alpha-amylase producing organism (ATCC 39,705). The final enzyme product, after fermentation and processing, contained less than 1.5 grams of the alpha-amylase protein per liter, but due to higher specific activity, exhibited an order of magnitude advantage over other commercial preparations during starch hydrolysis.

In the second case (EBS, 1986), genetic material containing the *B. megaterium* alpha-amylase gene was transformed into *B. subtilis*. The transformant produced *B. megaterium* amylase. After fermentation and processing, the final product contained 23 grams of the amylase protein per liter.

Chymosin
Chymosin (rennin) is a protease enzyme used in the clotting of milk to make cheese. The traditional form is obtained from calves' stomachs, although other substitute enzymes are also used. One of the first and most publicized examples of the genetic engineering technology has been the attempts to clone and economically produce the calf chymosin. To be successful, one must:

- economically produce the properly folded protein that produces cheese as good as that possible with native calf chymosin, and
- satisfy the regulatory concern.

Achieving this has been difficult so far, but very significant progress has been made. Among the leaders in this race have been Pfizer, Inc. and Genencor, Inc. Pfizer submitted to the FDA the first food additive petition involving a genetically engineered microorganism (Chemical Marketing Reporter, 1987). The company has made license arrangements with Celltech of the United Kingdom (Chemical Marketing Reporter, 1988), and Collaborative Research/Dow Chemical (Genetic Engineering Letter, 1988). Genencor has cloned and expressed chymosin in both bacterial and fungal hosts (Lawlis et al., 1987, Cullen and Berka, 1987, Cullen et al., 1987). Other companies with varying degrees of technology in this area include Codon, Allelix, Genex, and Gist Brocades (Pitcher, 1986, Genetic Technology News, 1986).

The chymosin gene has been expressed in virtually every host organism of conceivable practical interest. These include the bacterium *E. coli* (McCaman and Cummings, 1986; Taylor et al., 1986), the filamentous fungus *Aspergillus awamori* (Lawlis et al., 1987), and yeast *saccharomyces cerevisiae* (Smith et al., 1985). The yeast *Kluyveromyces* and the *bacillus* bacteria have also been pursued (Pitcher, 1986). It is only a matter of time before one or more of these efforts is successful in fully meeting the challenge.

Other starch enzymes
Further improvement in the economics of starch processing are possible with cloning and expression at economic yields of several of the other starch enzymes. More thermostable glucoamylase, better raw-starch-degrading enzymes, lower cost beta-amylase, alpha-D-gluco-

sidase, and cyclomaltodextrin D-glucotransferase are some of the potential targets (Kennedy et al., 1988).

The above examples of chymosin, alpha-amylase, and other enzymes illustrate how genetic engineering technology is being utilized to allow economical use of well defined but previously uneconomical enzymes.

New or Modified Enzymes for Known Applications

Perhaps of greater significance is the recent demonstration of tailored enzymes to achieve targeted enzyme activities. This ability has also been described as protein engineering, enzyme engineering, and site-directed mutagenesis. The examples of subtilisin, glucose isomerase, and lipase will be considered here.

Subtilisin

While a major use of the protease enzymes is in the detergent industry, proteolytic enzymes are also used in the food industry. Applications include meat tenderization, modification of dough in baking, improvement of flavor, texture, and color, and improved process efficiency.

Subtilisin is a bacterial protease, a serine endopeptidase secreted by a vareity of *Bacillus* species. This enzyme has been widely studied for genetic modification and its three-dimensional structure has recently been resolved to 1.8°A (Bott, et al., 1988).

Scientists at the Imperial College, London, UK (Russell and Fersht, 1987), and at Genencor and Genentech (Wells and Estell, 1988) have shown that the pH profile of subtilisin can be engineered. They hypothesized that by changing the surface charge of the enzyme its pH profile could be altered. Substitution at positions near one of the catalytic sites (Histidine 64) to replace negatively charged amino acids should lead to a decrease in the pK_a of the active His64. In the most dramatic case, the Imperial College scientists achieved a shift of one pH unit (at low ionic strength) with a double mutant which had its Asp99 replaced with Lys and Glu156 with Lys. As an illustration of the complexity of this, substitution of Met222 with Lys gave unexpected results suggesting a combination of steric and electrostatic affects may be involved (Graycar, 1988; Fig. 1).

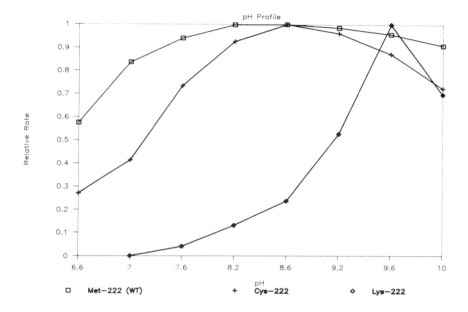

Figure 1. Shifting the pH profile of an enzyme by genetic modification. The surface charge of the enzyme (subtilisin) was altered by site-specific mutagenesis leading to substitutions at the amino acid site 222 of the enzyme (Reproduced with permission from Graycar, 1988). WT: wild-type enzyme.

Site-directed mutagensis has also been utilized to study the susceptibility of subtilisin to chemical oxidation. Subtilisin has a methionine residue at position 222 which was hypothesized to be responsible for its high susceptibility to oxidants such as hydrogen peroxide. Because of uncertainties in predicting which amino acid would be the best substitution, Estell et al. (1985) achieved the substitution of the methionine residue with all the other possible 19 amino acids. All 19 variant enzymes had activity. Figure 2 shows the dramatic improvement when ser222 or ala222 were present in place of the met222.

Subtilisin proteases are active against a wide variety of substrates, but the reaction rates vary a great deal. The kinetics depend on the specific subtilisin-substrate pair. Fig. 3 shows the *Bacillus amyloliquefaciens* and *Bacillus licheniformis* substilisins differ by a factor of 10-50 in catalytic efficiency depending on the substrate. (Wells and Estell,

1988). Although the enzymes have 86 amino acid differences out of a total of about 275, only three substitutions could modify the *B. amyloliquefaciens* subtilisin to have a similar catalytic efficiency as the *B. licheniformis* enzyme for several of the substrates studied.

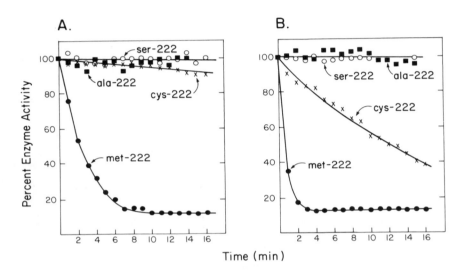

Figure 2. Engineering an enzyme to be resistant to chemical oxidation. Time course of the effect of 0.1 M H_2O_2 (A) or 1.0 M H_2O_2 (B) on the activity of purified subtilisin variants (Reproduced with permission from Estell et al. 1985. Copyright 1985 Amer. Soc. for Biochem. and Mol. Biol.). Wild type enzyme is indicated as met-222. Enzymes were incubated in fresh H_2O_2 and 0.1 M sodium borate for the times indicated. Residual activity is expressed as a percent of a non-treated control.

Further work in this area has illustrated the complexity involved. Scientists are discovering ways of examining the contribution of different chemical binding forces such as steric, hydrophobic, and electrostatic interactions on substrate specificity and catalysis. Fig. 4 (Wells et al., 1987) shows the effect of net charge in the substrate binding site on the catalytic efficiency. The net charge was varied by making substitutions at the 156 and 166 positions. In the extreme case, a 4,000 fold change in catalytic efficiency was achieved.

FOOD QUALITY • 295

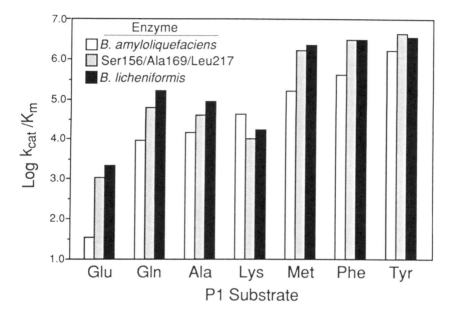

Figure 3. Altering the substrate specificity of an enzyme. Specificity profiles for the subtilisin from *B. amyloliquefaciens*, its variant subtilisin derived by substitutions at three positions (Gln156Ser, Gly169Ala, Tyr217Leu), and the subtilisin from *B. licheniformis* (Reproduced with permission from Wells and Estell, 1988. Copyright 1988 Elsevier Science Publications). Overall catalytic efficiency of the three enzymes is shown for several substrates.

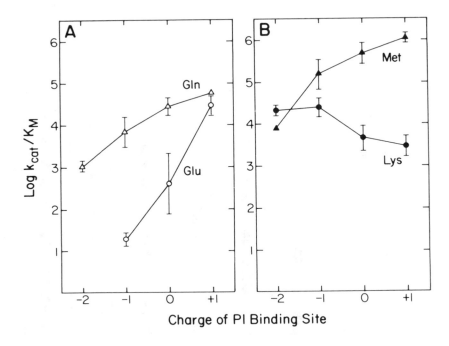

Figure 4. Designing substrate specificity by protein engineering of electrostatic interactions between an enzyme and its substrate. The charge in the enzyme-substrate binding site was changed by substitutions at positions 156 and 166 of the subtilisin enzyme (Reproduced with permission from Wells et al., 1987). Overall catalytic efficiencies for four substrates differing in their P1 amino acid (glutamine, glutamate, methionine, and lysine) are shown.

This efficacy to alter subtilisin goes beyond its hydrolytic activities. In a study to model a transesterification reaction (Poulose et al., 1987), the hydrolysis and methanolysis of tryosine ethyl ester was examined. The researchers chose position 222, an amino acid adjacent to the active 221 serine residue, for site-directed mutagensis because the 221 serine reacts with the nucleophile upon deacylation. Table 2 shows a four-fold improvement in the methanolysis to hydrolysis ratio by substituting phenylalanine for methionine.

Table 2. Transesterification by subtilisin variants

	Hydrolysis (mM)	Methanolysis (mM)	Methanolysis/ Hydrolysis
Wild type	8.87	6.38	0.72
M 22 F	2.49	7.86	3.16
M 222 Q	9.62	1.38	0.14

Reproduced with permission from Poulose et al. 1987.

Thus, although the specific changes in enzymes are difficult to design for now, the scientific knowhow is rapidly building. While the complexity may limit the ability to predict the exact effects of altering a few amino acids, significant changes in enzyme performance are unmistakably achievable.

Lipase
The lipases are among the more versatile group of food enzymes. Not only are the lipases meant for cleavage of fatty acids from their glyceride backbones, they can also be used for interesterification, transesterification, ester synsthesis, peptide synsthesis, biosurfactant production, and resolution of racemic mixtures to produce optically active compounds (Gillis, 1988). A lipase-based process to convert the relatively low cost palm oil mid-fraction to a more valuable cocoa butter equivalent is under development by Unilever (Righelato and Rodgers, 1987). The selectivity of some of the commercially available extracellular lipases is shown in Table 3 (Sonnet, 1988). This variety of their potential uses makes the lipase an attractive candidate for genetic modification.

Table 3. Positonal and fatty acid selectivities of some extracellular microbial lipases

Organism	Mfr.code	P selectivity	F A selectivity
Candida rugosa	Enzeco	Non-	18 (cis-Δ9)
Aspergillus niger	Amano-K	Non-	10,12
Geotrichum candidum	Amano-GC4	1,3-	18(cis-Δ9)
Aspergillus niger	Amano-AP	1,3-	18(cis-Δ9)
Mucor Miehei	Novozyme 225	1,3-	12
Rhizopus arrhizue	Gist Brocades	1,3-	8,10

Reproduced with permission from Sonnet, 1988.
Copyright 1988 Amer. Oil Chem. Soc.

Genetic modification of lipases, however, is more difficult because unlike substilisin, the crystal structure for lipase is not yet available. In one of the very few reports available, Poulose et al. (1987) studied a *Pseudomonas* lipase for modification of the enzyme for the hydrolysis and methanolysis of trioctanoin. Through a series of experiments, they selected positon 127 on the enzyme for modification and succeeded in altering the nucleophile specificity of the enzyme (Table 4).

Table 4. Transesterification/hydrolysis ratios of lipase variants

Enzyme	Methanolysis/Hydrolysis
Wild type	0.77
CYS 127	0.88
THR 127	1.35

Reproduced with permission from Poulose et al. 1987.

Glucose Isomerase

Researchers at the Imperial College, London, UK, have targeted a glucose isomerase enzyme for protein engineering (Hartley et al., 1987). They are interested in the *Arthrobacter* isomerae enzyme originally discovered by the Reynolds Tobacco Company. The initial attempts at cloning the isomerase gene have met complex problems as is often the case with pioneering work.

New/Modified Enzymes for New/Unknown Applications

Perhaps the greatest potential in the above areas lies in extending the use of enzymes to new and imaginative applications. In this concluding section, a few remarks will be made on potential applications and expected challenges.

At the risk of oversimplifying a very complex process, three things are required to commercialize a new food enzyme area:

- A legitimate application with sufficient market potential

- success in technology and market development efforts to economically meet that application, and

- success in satisfying, within the constraints posed by the first two factors, regulatory concerns involved.

The last of these may not be as obvious as the first two but nevertheless needs addressing. In the United States, the FDA regulates food enzymes under the Food, Drug and Cosmetic Act (FDC). The FDC states, in a simplified summary, that anything that is added to food or that affects the characteristics of a food is a food additive unless it is GRAS or approved for use prior to 1958 (Fordham and Block, 1987). The GRAS petitions accepted by the FDA from CPC (1984), EBS (1986), and Pfizer (Chemical Marketing Reporter, 1987) show that enzymes produced by the recombinant DNA technology may be GRAS.

Guidelines for assessing the safety of enzymes meant for food processing have been proposed (Pariza and Foster, 1983). Table 5 summarizes the safety program followed by Novo, Inc. for their immobilized lipase enzyme from *Mucor miehei*. In many new food applications with new enzymes, such scientific evidence may be required to obtain affirmation of GRAS status.

Table 5. Safety program for lipozyme™

I. Safety of the production strain
- low virulence, shown by intravenous inoculation of spores in mice
- no antibiotic substances
- no mycotoxins
- absent from enzyme preparation

II. Toxicoloty program on lipase preparation (40 000 LU[a]/g)
- acute toxicity, rats : LD-50 > 10g/kg
- 28 days feeding, rats : no toxic effect (0.5 g/kg)
- 14 days feeding, dogs : no toxic effect (6 g/kg)
- 90 days feeding, rats : no toxic effect (1 g/kg)
- 90 days feeding, dogs : no toxic effect (5 g/kg)
- mutagenicity, Salmonella : not mutagenic (133 LU/ml)
- cytogenicity, rats : no efffect on chromosomes (5 g/kg)
- teratogenicity, rats : no teratogenic effect (3.3 g/kg)

III. Physical-chemical stability of Lipozyme
- checks for leakage from macroporous anion exchange resin
- check for leakage of enzyme from Lipozyme

[a]LU = Lipase units.
Reproduced with permission from Fordham and Block, 1987.
Copyright 1987 Soc. for Ind. Microbiol.

The pursuit of new applications and the use of genetic engineering or protein engineering technologies for economical exploitation is likely to be an iterative process. The number of potential applications for food enzymes is probably only limited by our imagination. Rattray (1984), Righelato and Rodgers (1987), and Klacik (1988) list opportunities in the areas of oils and fats, carbohydrates, and reactions that require co-factors. Some of these are also listed in Table 6. A few examples will be mentioned here.

Cell wall fiber degrading enzymes are known to be effective in the extraction of oils from oilseeds (Klacik, 1988). If the appropriate enzyme activities could be defined and produced economically, then safe and competitive alternatives to solvent based oil extractions may get commercialized.

Table 6. Some desirable targets for use of enzymes

Substance	Target Area
Vegetable Oils	Increase in the degree of unsaturation
	Biohydrogenation without undesirable fatty acid intermediates
Glycerides	Controlled hydrolysis from triglycerides to make mono- and diglycerides and fatty acids
	Controlled synthesis of glycerides from glycerol and fatty acids
	Controlled acyl exchange for the modification of triglycerides
Butterfat	Controlled interesterification to produce cocoa butter-like fat from cheaper oils and fatty acids
Lecithin	Controlled recovery from soybean oil to achieve particular biosurfactant activity
Pectins	Increase in the extent of methylation
	Removal of acetate groups without removing methoxyl groups
Guar gum	Introduction or removal of alpha-D-galactose
Locust bean gum	side chains
Alginic acid	Modification in the ratio of mannuronic and fuluronic acid residues

Derived from Rattray (1984) and Righelato and Rodgers (1987).

Given the appropriate lipase selectivities, enzymatic routes to ester synthesis can be achieved. Conversely, the hydrolytic specificity of lipases may be exploited for making food grade emulsifiers from triglycerides (Rattray, 1984). If cost, thermostability, and specificity targets for lipase activities could be met, the steam hydrolysis process for splitting fats could be replaced with a better process (Brady et al., 1988). Modifications in the degree of saturation or unsaturation in oils and fats are opportunities on nutritional and economic grounds.

New applications for new or modified enzymes may also be possible in the modification of polysaccharides which are used in the food industry as gelling agents. Use of enzyme activities to introduce methyl groups in pectins, or to add or remove side chains in gums, or to interconvert mannuronic-guluronic acid residues in alginates are some areas worthy of investigation (Righelato and Rodgers, 1987). Other possibilities include improvement of nutritional properties of foods, and modification of proteins to tailor flavor, texture, emulsification, viscosity etc., type properties (Righelato and Rodgers, 1987).

On the other hand, very good progress also continues on the use of genetic engineering technologies. Progress on starch enzymes, chymosin, substilisin, and lipase has been reviewed here. As reported by Mead et al. (1986), at least 19 cellulase or amylase genes have been cloned, including specific cellulase components (Chen et al., 1987). Heterologous gene expression systems have been developed for *Escherichia coli*, *Saccharomyces cerevisiae*, and *Aspergillus nidulans* (Cullen and Berka, 1987). Plant enzymes used in the food industry, such as papain and beta amylase, are being studied for potential improvements (Smith et al., 1987). Cost effective use of anticholesterol enzymes in dairy processing is seen to have a potential of over $100 million (Genetic Technology News, 1988). Researchers at Iowa State University are examining the possibility of modifying beta-galactosidase by fusion of acidic and basic polypeptides to improve the efficiency of post-fermentation processing.

All this adds up to bright prospects for the technology of genetic modification of food enzymes. Only the degree of potential and the rate of progress are unknown.

Acknowledgements

I would like to thank W.H. Pitcher, T.P. Graycar, A.J. Poulose, J.A. Wells and D.A. Estell for their generous help and support, and for the use of their data.

References

Bott, R., Ultsch, M., Kossiakoff, A., Graycar, T., Katz, B. and Power, S. 1988. The three-dimensional structure of *Bacillus amyloliquefaciens* at 1.8°A and an analysis of the structural consequences of perixode inactivation. J. Biol. Chem. 263(16): 7895-7906.

Brady, C., Metcalfe, L., Slaboszewski, D. and Frank, D. 1988. Lipase immobilized on a hydrophobic, microporous support for they hydrolysis of fats. J. Amer. Oil Chem. Soc. Vol 65(6): 917-921.
Chemical Marketing Reporter 1988. Celltech signs rennin pact; takes on partner for lipase. February 8.
Chemical Marketing Reporter 1987. Biotech for cheese makers. December 7: 9.
Chen, C.M., Ward, M., Wilson, L., Sumner, L. and Shoemaker, S. 1987. Toward improved cellulases: targeted modifications of *Trichoderma reesei* exo-cellobiohydrolase II using site specific mutagenesis. Presented at the 194th Meeting of the American Chemical Society, August 30- September 4, at New Orleans, Louisiana.
CPC International, Inc. 1984. (Submitted). Affirmation of GRAS status of alpha-amylase enzyme from *Bacillus subtilis*. GRASP 4G0293.
Cullen, D. and Berka, R.M. 1987. Molecular genetics of commercially important filamentous fungi. Chimicaoggi- gennaio. February: 57-59.
Cullen, D., Gray, G.L., Wilson, L.J., Hayenga, K.J., Lamsa, M.H., Rey, M.W., Norton, S. and Berka, R.M. 1987. Controlled expression and secretion of bovine chymosin in Aspergillus nidulans. Biotechnology. Vol. 5 April: 369-376.
Enzyme Biosystems, Inc. 1986. (submitted). A petition for the affirmation of the GRAS status of amylase of *Bacillus megaterium* derived from *Bacillus subtilis*. GRASP 7G0328.
Estell, D.A., Graycar, T.P. and Wells, J.A. 1985. Engineering an enzyme by site-directed mutagenesis to be resistant to chemical oxidation. J. Biol. Chem. 260(11): 6518-6521.
European Chemical News 1986. Enzyme demand growth shows signs of slowing down. March 31: 11-12.
Fordham, J.R. and Block, N.H. 1987. Regulatory issues of enzyme technology. Dev. in Indust. Microbiol. 28: 25-31.
Genetic Engineering Letter 1988. Dairy products: Genetic engineering helps. April: 6-7.
Genetic Technology News 1986. Market for recombinant enzymes for the food industry. August: 6-8.
Gillis, A. 1988. Research discovers new roles for lipases. J. Amer. Oil Chemists Soc. 65(6): 846-850.

Graycar, T.P. 1988. Protein engineering of subtilisin. Presented at the 32nd Annual Wind River Conference on Genetic Exchange, June 5-9 at Estes Park, Colorado.

Hartley, B.S., Anderson, T. and Shaw, P-C. 1987. Protein engineering of food enzymes. In Chemical Aspects of Food Enzymes. A.T. Andrews (ed.) Royal Society of Chemistry Special Publication No. 63: 120-136.

Kennedy, J.F., Cabalda, V.M. and White, C.A. Enzymic starch utilization and genetic engineering. Trends in Biotechnol. Vol. 6(8): 184-189.

Klacik, M.A. 1988. Enzymes in food processing. Chem. Eng. Prog. 84(5): 25-29.

Lawlis, V.B., Hayenga, K.J. and Heinsohn, H. 1987. Characterization and model studies on the recovery of chymosin produced by filamentous fungi e.g. *A. Awamori*. Presented at teh 194th Meeting of the American Chemical Society, August 30-September 4, at New Orleans, Louisiana.

Marie-Henrietta, D., Gunther, H. and de Troostembergh, J. 1985. Process of enzymatic conversion. European Patent Application No. 8530 3493.2 Publication No. 0 164 933 Date of Publication 18.12.85 Bull. 85/51.

McCaman, M.T. and Cummings, D.B. 1986. A mutated bovine prochymosin zymogen can be activated without proteolytic processing at low pH. J. Biol. Chem. 261(33): 15345-15348.

Meade, J.H., White, T.J., Shoemaker, S.P., Gelfland, D.H., Chang, S. and Innis, M.A. 1986. Molecular cloning of carbohydrases for the food industry. In "Impact of Biotechnology on Food Production and Processing," D. Knorr (ed.). Marcel Dekker, Inc. New York.

Pariza, M.W. and Foster, E.M. Determining the safety of enzymes used in food processing. J. Food Protection. Vol. 46(5): 453-468.

Pitcher, W.H. 1986. Genetic Modification of Enzymes Used in Food Processing. Food Technol. 40(10): 62-63 & 69.

Poulose, A.J., Pepsin, M.J., Estell, D.A. and Arbige, M.V. 1987. Improved transesterification through enzyme engineering. Presented at the 78th Meeting of the American Oil Chemists Society, May 17-21, at New Orleans, Louisiana.

Rattray, J.B.M. 1984. Biotechnology and the fats and oils industry - an overview. J. Amer. Oil Chem. Soc. Vol. 61(11)1701-1712.

Reichelt, J.R. 1983. Strach. In "Industrial Enzymology," (T. Godfrey and J. Reichelt, eds). The Nature Press, New York, NY.

Righelato, R.C. and Rodgers, P.B. 1987. Food enzymes: industrial potential and scientific challenges. In Chemical Aspects of Food Enzymes. A.T. Andrews (ed.) Royal Society of Chemistry Special Publication No. 63: 271-288.

Russel, A.J. and Fersht, A.R. 1987. Rational modification of enzyme catalysis by engineering surface charge. Nature 328: 496-500.

Scott, D. 1980. Industrial enzymes. In Kirk-Othmer Encyclopedia of Chemical Technology. 3rd Edition. John Wiley and Sons, New York, Vol. 9: 204-205.

Smith, H., McKee, R.A. and Praekelt, U. 1987. The molecular biology of plant thiol proteinases. In Chemical Aspects of Food Enzymes. A.T. Andrews (ed.) Royal Society of Chemistry Special Publication No. 63: 197-207.

Smith, R.A., Ducan, M.J. and Moir, D.T. 1985. Heterologous protein secretion from yeast. Science 229: 1219-1224.

Sonnet, P.E. 1988. Lipase Selectivities. J. Amer. Oil Chemists Soc. 65(6): 900-904.

Tamuri, M., Kanno, M. and Ishii, Y. 1981. Heat and acid-stable alpha-amylase enzymes and processes for producing the same. U.S. Patent 4,284,722. August 18.

Taylor, G., Olbrich, R., Gardiner, S.A.M., Gray, D.J., Marston, F.A.O., Hoare, M. and Fish, N.M. 1986. Physical properties of protein inclusion bodies. Presented at the 192nd ACS National Meeting, September 7-12, at Anaheim, California.

Wells, J.A. and Estell, D.A. 1988. Subtilisin - an enzyme designed to be engineered. To be published in Trends In Biol. Sci.

Wells, J.A., Powers, D.B., Bott, R.R., Graycar, T.P. and Estell, D.A. 1987. Designing substrate specificity by protein engineering of electrostatic interactions. Proc. Natl. Acad. Sci. 84: 1219-1223.

Impact of Biotechnology on Vegetable Processing

N. D. Addy
The ADDCO Group
146 Gold Creek Circle
Folsom, CA 95630, USA

The food processing industry has begun to recognize and utilize the valuable tools of biotechnology to improve the quality of processed vegetable crops. The three major vegetable crops gaining deserved biotech R&D attention are tomato, potato, and cole crops. Cucurbits, onions and garlic represent other minor high-value processed crops. Targets for key crop improvement are increased field yields, enhanced processing yields, improved processing quality, increased nutritional value and overall consumer value.

This paper describes the biotechnological developments which are allowing the food processing industry to take advantage of new genetically improved vegetable crops suitable for enhancing existing food products and the development of new cultivars of processed vegetables for the international marketplace.

Introduction

The development and marketing of new improved varieties of vegetables for the food processing industry has historically occupied an important place in the commercial efforts of many agricultural seed companies. In the field of plant agriculture, biotechnologists are faced with the same problems that plant breeders have struggled with for years. Among others, these include yield enhancement, enhanced crop processing quality, and improved tolerance to a variety of insect and disease organisms. The newly emerging field of agricultural biotechnology will not alter the problems facing the food processing industry—it simply represents a new approach to solving them.

Agricultural Biotechnology Industry

Agricultural biotechnology has been defined in many different ways, depending on one's interpretation of biotechnology. I prefer to visualize agricultural biotechnology as a tool box containing a diverse collection of tools. Each tool can be utilized separately or together in the process of genetically modifying existing crop varieties. This creative building activity can lead to new unique plant varieties which can possess commercially important characteristics. The collection of tools in the tool box relies on techniques used in varied disciplines such as physiology, cell biology, molecular genetics, plant breeding and biochemistry. The challenge facing the food industry is how to best integrate and use the tools of biotechnology to produce commercially viable products.

The genetic manipulation of a vegetable crop variety using the tools of biotechnology is analogous to the process of cutting a rough diamond to form a valuable precious stone. Each new cut surface or facet of the stone must be produced with utmost care using the proper tools in the hands of a skilled craftsman. One miscalculation by the diamond cutter could significantly reduce the value of the finished stone. Each facet must fit the pattern of the whole stone thus forming a system composed of the setting, the overall size of the stone and the various cut surfaces that enhance its total value.

The use of plant biotechnology to improve crop varieties is nothing more than a process of determining 1) what trait needs to be changed genetically, 2) where that trait resides in the genetic structure, and 3) changing or substituting the appropriate genetic information to produce the desired result by utilizing the available biotech tools. As with the cutting of a rough diamond, the biotechnologist risks miscalculating the results of genetic change. This can spell varying degrees of trial and error in developing an improved vegetable variety that answers the needs of the grower and processor. Unlike the case of the diamond cutter, however, the biotechnologist can start over with the same original working germplasm and the process can be repeated using slightly different approachs for genetic manipulation.

Even though highly specialized tools are being added each year to the tool box, the rate of new product development or improvements in food crops remains very sluggish and disappointingly slow. Of paramount importance to this process is time and money. Lengthy R & D development periods have hampered the plant biotechnology industries effort to produce quality products. Moses et al. (1988) have pointed out that the benefits of agricultural biotechnology will largely remain unrealized until funding programs accommodate the two important qualities of biotechnology research. First, they must recognize that biotechnology is not a scientific discipline itself, but rather a collection of tools as described previously. Second, that the tools of biotechnology, for the most part, operate at the molecular level. Their application requires a large initial investment in basic research, focused on isolating and moving the genes that determine important traits. Additionally, the companies involved in biotechnology must have access to financial resources adequate to maintain a full range of technologies and adequate marketing skills to manage and direct the creative use of the tools in developing products of value.

Food Industry Needs

When evaluating the needs for new or improved vegetable crops for the food processing industry, one should view the supply of quality raw vegetable products to the industry as an inter-linked process or system. The first link in the chain is the supplier of the germplasm. Historically, the supplier has been a traditional seed company field producing true seed varieties. More recently seed companies are using the tools of biotechnology to supplement their field breeding programs. New suppliers of germplasm are agricultural biotechnology companies which have emerged since 1980.

The second link is composed of growers who plant, harvest and market the food crop destined for the food processing industry and, ultimately, the consumer. The last link in this loop is the vegetable processor with a set of expectations and requirements for raw vegetable product quality. The desired processing traits are essential for producing a commercially viable vegetable product acceptable to the consumer and profitable to the processor. These three functional parts of the food processing industry can form a self supportive closed loop with each in the position to share in the revenues derived from new and improved processed vegetable products.

Crop innovations presents a formidable barrier for market entry of new agricultural biotechnology seed companies. Strategies for succeeding in an already risky market must be developed and analyzed carefully. New roles must be played by all the participants who expect to profit from crop improvements. Even then, only the strongest biotechnology companies will have a reasonable chance of succeeding.

Processed Vegetable Crops

A wide range of vegetables crops is processed by the food industry into consumable food products. These include tomatoes, potatoes, sweet corn, legumes, cole crops, cucurbits, onion and garlic. However, for discussion in this paper, I will limit my remarks to tomatoes and potatoes. These two food crops represent over 60% of the vegetables processed world wide and 73% in the U.S. Table 1 lists vegetables used in the food processing industry). Ratafia and Purinton (1988) identified tomatoes and potatoes as candidates for genetic engineering and commercialization in 1989.

Processed tomatoes constitute a very important food crop both in the United States and worldwide. In 1986, production in the U.S. amounted to over 7,000,000 tons with a market value of $473,000,000. World production exceeds 47,000,000 metric tons.

Table 1. Principal vegetable crops utilized for processing in the United States

Vegetable	Acres harvested	Raw product statistics Tons processed	Value per ton($)
Cucumber	110,300	637,400	177.00
Pea	272,400	415,000	226.00
Snap Bean	190,600	609,400	160.00
Sweet Corn	413,000	1,012,240	59.00
Tomato	252,100	7,393,300	64.00
Potato	501,600	8,660,000	75.00
Asparagus	96,200	42,100	928.00
Broccoli	108,900	127,700	369.00
Carrots	86,000	360,100	63.00
Onions	100,000	2,138,750	175.00
Garlic	25,000	207,493	371.00
Others	10,000	200,000	150.00
Total	2,166,100	21,803,483	
Average dollar value per ton			$235.00

The potato is one of the world's most valuable food crops. It is grown in more countries than any other crop except maize. Its volume of production ranks fourth in the world after rice, wheat, and maize. In edible energy and yield of protein per hectare per day, the potato is near the top of the list of major world food crops. In recent years, the growth rate of potato production developing countries has exceeded that of most other food crops.

Tomatoes

Tomatoes bred and grown for the processing industry are generally the result of a close linkage between the seed producer, the grower, and the processor. The processor tends to be the controlling factor in this three link chain. New tomato varieties produced by the various seed companies gain acceptance in the market places only after approval by the processors and contract growers. The process is slow and methodical and, in some cases, involves heavy-handed politics. Even though the process lacks perfection it does provide direction to the seed companies in their efforts to develop new and improved varieties of tomatoes which will address the needs of the industry.

Using traditional field breeding techniques to generate new processing tomatoes requires five to seven years of development time before these new varieties can compete in the market place. Biotechnology has the potential to reduce this development time to two or three years. Specific genetic targets include increased solids, content, controlled ripening, improved color and enhanced yield. Once a plant regeneration is perfected for a germplasm line, numerous genetic selections can be produced and passed on to the field breeding staff for evaluation. Time is very valuable in this development process. Companies where the biotechnology staff works closely with their field breeding staff are generating new and improved germplasm in much shorter time frames.

Since tomatoes are currently produced by true seed, large volumes of a new seed variety can be made available to contract growers. Therefore, large volumes of the new crop selections are produced for the processing industry in a relatively short period of time. This is also true of hybrid seed varieties which takes longer to cross pollinate and produce large seed volumes than does the open pollinated varieties. As the seed companies increase their dependence on biotechnology to develop new varieties, more of the new seed cultivars will be hybrids. Hybrids derived through biotechnology offer the potential for hybrid vigor and proprietary protection.

Given significant breakthroughs in tissue culture, process tomato varieties may be clonally produced as transplants in the near future and thus, bypass production of true seed all together. Although clonal propagation is normal in potato production, the cost constraints of tissue culture prevent this tissue culture approach from being used today in the tomato industry.

In the final analysis, a new and improved processing tomato variety must compete with the current spectrum of tomato varieties on the basis of yield and crop quality. The growers are in a position to share only a small portion of the perceived value from crop improvement through biotechnology. The growers' willingness to share the increased cost of the improvement will depend on the food processors' willingness to recognize the value of crop improvements and participate in a value-sharing process. The marketplace will ultimately place limits on premiums which can be realized from crop improvements through biotechnology.

Potatoes

The process of potato breeding and variety election is very similar to the breeding techniques used for tomato and other true seed-based vegetable crops. Potato field breeders make their zygotic, genetic crosses; collect and plant the resulting seed; and evaluate the resulting plants for the correct combination of valuable characteristics. However, potato derived from true seed crosses has a high level of heterozygosity, a high rate of pollen sterility, selection difficulties in the seedling stage, and a slow rate of growth.

The development of new potato varieties using standard field breeding techniques takes considerably longer than a similar breeding program for tomato. The main breeding objective is to increase yield while maintaining tuber quality. This could involve altering the date of plant maturity, increasing the growth rate, increasing the number of uniform and salable tubers per plant, and genetically introducing resistance to insects and microorganisms (Abel et al., 1986). Processing objectives include high solids content, low reducing sugars, resistance to discoloration during processing and improved tuber storage characteristics.

Except for the initial field breeding activities, the potato is a clonally propagated crop where the selected variety is maintained using a vegetative planting stock of sectioned potato tubers, each containing several sprout sites or "eyes." This method of field production is carried on without the aid of true seed and is analogous to vegetative clonal propagation. Essential to any commercial potato program is the production of quality seed potatoes free of viruses and other plant diseases. High quality, disease free propagation stock is readily available through clonal propagation of vegetable tissue under germ free conditions in tissue culture. These tissue culture nodal cuttings can then be rooted to form individual potato plantlets and transplanted directly to the field or in greenhouses to produce the first generation of small tubers generally known as minitubers.

The commercial application of micro-propagation techniques to the production of seed potatoes and other vegetable crops in tissue culture has been a recent phenomenon (Roca, 1985). In North America, the tissue culture propagation has been adopted by most of the states and provinces involved in the production of early generation seed potatoes. More recently, a number of private biotechnology based companies have begun the commercial production of seed potato products.

Given todays production technology, the number of annual field generations necessary to supply the quantity of seed potatoes for commercial growers has been reduced to an average of six years. Most of this initial (nuclear) seed planted by the seed growers is derived from tissue cultures available throughout North America. Similar tissue culture based, seed potato programs exist throughout Europe and the Pacific Rim countries. With the exception of several private companies, the bulk of this initial planting stock is produced and supplied to the growers by various state and provincial organizations. These programs tend to be expensive to operate and maintain. There is a tradeoff between costs and total number of generations grown in the field. A key objective of the private biotech companies is to produce large volumes of (nuclear) tubers in one generation at a much reduced cost. This would further reduce the number of field generations necessary to produce the commercial potatoes for the processing industry and fresh market.

Because the potato is produced via clonal propagation systems, it presents an excellent opportunity for the agricultural biotechnology groups to rapidly introduce new genetic material into the variety evaluation system (Addy and Stuart, 1986). The quality of potato varieties should continue to improve rapidly with the availability of new laboratory based molecular genetics that accelerate the identification of new traits for introduction and evaluation.

Comprehensive integration of field and laboratory plant breeding as moving the development process for new varieties forward at a rapid pace. Automated tissue culture systems are also close at hand and several patented systems are available in Europe, Israel and Australia (Rich, 1988). As these new automated innovations are introduced to tissue culture technology, they should have a direct impact on reducing the production cost of clonal material. These robotic breakthroughs will also have a direct effect on the cost of producing seed potato tubers.

Discussion

The time it takes to gain acceptance of an innovative crop in the marketplace can be the most significant constraint in the high-tech development approaches using new technology. In addition to the time it takes to successfully carry out genetic manipulation of a plant variety, the new germplasm must then be subjected to rigorous cultivar testing, regulatory and registration evaluations, and market exposure. For example, the current time frame for development of California processing tomato varieties is three to four years of basic field breeding and selection. An additional three to five years of expanded testing is required by the processing industry and the grower/ag extension complex to establish performance and accept the new variety in the marketplace.

The vegetable industry as a whole and the potato/tomato processing industry specifically, should expect more new varieties to be developed, introduced, and protected by private biotechnology companies (Addy, 1988). This prospect could lead to fewer varieties in the market place at any one time, but could also contribute to major advances in the control of diseases and insect problems plaguing growers today. It could also result in a greater emphasis being placed on the development of varieties possessing important processing traits. Since the new biotechnologies will allow more rapid introduction of improved varieties, targeted to meet the specific needs of the processing industry, we should see a shortening of each individual variety's total life span. For that reason, the industry could realize an increase in the number of developed varieties over a period of several decades, but a significant reduction in the number of major varieties being used at any one time.

The potential for creating new improved vegetable crops which are more suitable to the needs of the food processing industry brings with it the prospect of major changes in the agricultural production system. Growers are already being effected by rapid changes in the available germplasm of tomato and potato as well as other vegetable crops. Rapid change in any industry can be both stimulating and profitable to the participants. However, each of the participants must recognize the risks and benefits associated with the rapid change and be able to adapt to those changes.

The developing Agricultural Biotechnology industry, as it is successful in establishing profitable commercial operations, will have an ongoing responsibility of educating the food processing industry to the usefulness and value of their products and technology. As more and more of the new genetic engineering technologies and resulting vegetable products are introduced, an inevitable change in the agricultural and food processing industries will occur. Change in established industries is normally viewed with reservation and suspicion. To make these technological changes more palatable and acceptable to the food industry as a whole, the agricultural biotechnology based firms must accept their share of the responsibility for facilitating the orderly restructuring of the industry.

Summary

The biotechnology companies that are involved in developing new and improved vegetable varieties for the food processing industry are faced with enormous research and development costs long before revenues can be expected. These long developmental periods are further complicated by recognition that newly introduced varieties normally remain viable in the marketplace for only five to ten years before they are replaced by newer varieties. This gives little time for the industry to recover development costs along with a reasonable profit.

A need exists for establishing close relationships between the biotechnology companies and the food processing industry. A system of support between the biotechnology company, the growers of the vegetable product, and the vegetable processor could generate an acceptable revenue return to the biotechnology company from both the grower and the food processor. Again, this approach would depend on the beneficial linkage between the biotechnology company, the grower, and the processor. Through this linkage each party could profit from the revenue flow and the shared risk in developing new vegetable cultivars for the food industry.

For the emerging agricultural biotechnology companies involved in the development of new vegetable cultivars to be competitive in the traditionally conservative food industry, they must concentrate on the following three approaches to the production and sale of their products or technologies: 1) consistent production of high quality products, 2) development of strong service support of their products or technologies, and 3) constant awareness of the need to deliver value to the marketplace. All three of these product development activities are linked and contribute collectively to the expected success of each commercial venture.

References

Abel, P.P., Nelsen, R.S., De, B., Hoffman, N., Rogers, S.G., Fraley, R.T. and Beachy, R.N. 1986. Delay of disease development in transgenic plants that express tobacco mosaic virus coat protein gene. Science 232: 738.

Addy, N.D. 1988. Opportunities and challenges for private industry:Symposium on Applications of Tissue Culture and Micropropagation Techniques for Potato Production. American Potato Journal 65: 221.

Addy, N.D. and Stuart, D.A. 1986. Impact of biotechnology on vegetable processing. Food Tech. 40: 64.

McKell, C. 1987. Commercializing new products from plant biotechnology: problems and prospective. In: "Biotechnology: Perspectives, Policies, and Issues. pp. 188-206. University Presses of Florida, Gainesville, Florida.

Moses, P.B., Tavares, J.E. and Hess, C.E. 1988. Funding agriculture biotechnology research. Biotechnology 6: 144.

Ratafia, M. and Purinton, T. 1988. World agriculture markets. Biotechnology 6: 280.

Rich, R.P. 1988. Automated micropropagation - new system is leap forward. Agricultural Genetics Report 7(2): 197.

Roca, W.M. 1985. *In vitro* clonal propagation to eliminate crop diseases. In: "Biotechnology and International Agricultural Research." IRRI Press, Manila, Philippines.

Genetic Engineering of Lactic Starter Cultures

Larry L. McKay
Department of Food Science and Nutrition
University of Minnesota
1334 Eckles Avenue
St. Paul, MN 55108, USA

Application of biotechnology to dairy starter cultures has included classical genetic improvement and the development of gene cloning techniques essential for strain construction strategies. The latter encompasses establishing suitable transformation systems, constructing needed plasmid vectors for use in cloning, and the possible improvement of genes important in dairy fermentation processes. Plasmids have served as a focal point for genetic studies in dairy streptococci, since many code for properties vital in milk fermentation processes. Such properties include lactose fermentation, proteolytic activity, citrate utilization, exopolymer production, bacteriophage resistance, and production of inhibitory proteins. The plasmid linkage of these traits has facilitated genetic manipulation. Examples to be discussed include cloning of genes responsible for lactose utilization and inhibitory protein production, as well as isolating mutants to serve as delivery hosts for quick release of enzymes that might accelerate cheese ripening. These types of studies should lead to "optimized" lactic starter cultures.

Introduction

Interest in the genetics of dairy streptococci has rapidly expanded over the past 10 years. As stated by Gasson and Davies (1984), this is in part because the genetic technology now being developed has opened the way for industrial strain improvement programs in the dairy industry and because the genetic phenomena so far discovered provide a wealth of scientific information for the academic geneticist and molecular biologist. It is a young field of research, and many opportunities exist not only for the application of the genetic studies but also for providing fundamental information for an industrially important group of bacteria (Gasson and Davies, 1984). This paper provides a brief update on the genetics and application of biotechnology to lactic streptococci used in dairy fermentations. It is not intended to be a comprehensive review of the subject area, but rather emphasizes some of the work being conducted in our laboratory.

Plasmids

It is now well established that the mesophilic streptococci used in dairy fermentation processes harbor plasmids of diverse sizes (McKay, 1985). The average number of plasmids per cell ranges from 4 to 7, but the actual number may be from 1 to 2 to more than 12 per cell. Most of the plasmids are cryptic in that their function remains unknown. Some, however, have not been linked to identifiable metabolic characteristics (see Table 1).

Table 1. Metabolic functions that have been linked or tentatively linked to plasmid DNA in dairy streptococci

	Function	Selected References
1.	Sugar utilization	
	a. Lactose	de Vos and Simons, 1988; McKay, 1983; McKay, 1985
	b. Galactose	Crow et al., 1983
	c. Sucrose	LeBlanc et al., 1980
2.	Proteinase activity	Kok and Venema, 1988
3.	Citrate utilization	Kempler and McKay, 1981; Gasson and Davies, 1984
4.	Production of bacteriocins	Klaenhammer, 1988
5.	Nisin resistance independent of nisin production	McKay and Baldwin, 1984; Klaenhammer and Sanozky, 1985
6.	Resistance ot UV	Chopin et al., 1985
7.	Bacteriophage resistance mechanisms	Klaenhammer, 1987; Daly and Fitzgerald, 1987; Sanders, 1988
8.	Exopolymer production	Vedamuthu and Neville, 1986; von Wright and Tynkkynen, 1987; Neve et al., 1988

Gene Transfer Systems

To facilitate the manipulation of desired plasmid or chromosomally-linked genes for strain construction purposes, the development of gene transfer systems in dairy streptococci was required. The first to be described was transuction, in which a bacteriophage is involved in the genetic transfer process. Phage infection begins with adsorption of the phage on the bacterial cell surface. The DNA (viral, plasmid, or chromosomal) contained inside the phage head is then injected into the cytoplasm of the bacterial cell where it may become functional. Conjugation, the exchange of genetic material between a donor cell and a recipient cell as a result of physical contact between the two cell types, is the most common means of genetic exchange among dairy streptococci. Both transduction and conjugation have now been used to confirm that many of the metabolic traits vital to dairy fermentation processes (see Table 1) are plasmid coded in the mesophilic lactic streptococci. Transduction has also been used in the structural analysis and mapping of plasmid molecules from lactic streptococci and in stabilization of plasmid-linked traits such as lactose metabolism and proteinase activity. High-frequency conjugative plasmids from dairy streptococci have been observed which may be important in constructing strains for industrial purposes, since it will be necessary to have efficient systems for moving genes from one strain to another. It may also be possible to clone these high-frequency transfer regions onto other, non-conjugative plasmids, in order to transfer the plasmid into a desired strain. Alternatively, high-frequency conjugative plasmids might be developed into cloning vectors to dairy fermentation organisms. The genetic and molecular analysis of these conjugative plasmids, as well as the conjugative process itself, will be of considerable value for gene transfer programs. For an elegant discussion of transuction and conjugation processes in lactic streptococci, the reader is referred to the review by Fitzgerald and Gasson (1988).

Although the above gene transfer systems have aided in studying the genetics and plasmid biology of these organisms, the development of an efficient plasmid transformation system was vital for further genetic studies. Genetic transformation is the process by which DNA in the surrounding medium is taken up by the bacterial cell. A functional transformation system would facilitate genetic analysis and strain modification, as well as allowing application of recombinant DNA technology for strain improvement programs. A polyethylene glycol (PEG)-induced transformation of lactic streptococcal protoplasts using plasmid DNA has now been described from several laboratories. Sanders and Nicholson (1987) also described a procedure for the plasmid transformation of whole cells of *S. lactis* using PEG. A third process that has been introduced for getting free DNA into dairy streptococci cells is electroporation. This consists of exposing cells and DNA to high-voltage discharges in order to get uptake of the DNA by the cells. This process, as well as that of protoplast transformation, has recently been reviewed by Mercenier and Chassy (1988). Methods for the genetic manipulation of dairy starter cultures have been developed to the point that it is now possible to clone and express homologous and heterologous genes in these bacteria (de Vos, 1987).

Lactose Metabolism

In order to grow and function in milk, strains of *Streptococcus lactis* and *S. cremoris* must be able to ferment lactose. It is now known that the ability of these organisms to utilize lactose is due to the presence of a phosphoenolpyruvate dependent phosphotransferase system in which lactose is translocated into the cell as lactose-phosphate (McKay, 1985; de Vos and Simons, 1988). The specific enzymes involved are Enzyme II-Lac, Enzyme III-Lac, and phospho-β-galactosidase. Lactose-phosphate is then hydrolyzed to glucose and galactose-6-phosphate. It is also well established that the genes coding for these three Lac-specific enzymes are linked to plasmid DNA in these strains. Since the ability to ferment lactose is a plasmid-mediated trait in these bacteria, cloning the genes for lactose metabolism on a high copy number plasmid could increase the gene dosage and thus conceivably increase the rate of acid production. The cloning of these genes on a high copy number plasmid occurred naturally in a derivative of *S. lactis* (Anderson and McKay, 1984). Although the phospho-β-galactosidase activity was about twice that found in the strain containing the normal lactose plasmid, it had no effect on the rate of growth or rate of acid production when the cells were propagated in milk. This suggests that the lactic streptococci may already be optimized for lactose utilization and/or that the rate of lactose utilization is not the rate limiting step for acid production.

Unlike the mesophilic streptococci, *S. thermophilus* has received considerably less research attention. This organism hydrolyzes lactose by β-galactosidase to yield glucose and galactose. The β-galactosidase gene is assumed to be chromosomally linked, since strains with no detectable plasmids ferment lactose and exhibit β-galactosidase activity (Herman and McKay, 1985). The hydrolysis of lactose to its monosaccharides is of interest for several reasons. First, lactose is poorly digested by a large proportion of the world's population (Hourigan, 1984). Also, lactose has a low solubility in water which leads to problems in concentrating whey or with other food items containing whey, and lactose has a relatively low degree of sweetness (Smart et al., 1985). These problems, however, can be overcome to some extent by the hydrolysis of lactose to its monosaccharides which are more digestible and more soluble as well as sweeter than lactose (Smart et al., 1985). Investigators have thus searched for microorganisms capable of producing high levels of β-galactosidase (Wierzbicki and Kosibowski, 1973; Blankenship and Wells, 1974; Sorensen and Crisan, 1974; Rao and Datta, 1978). Although a number of sources of β-galactosidase now exist, some workers have suggested that *S. thermophilus* is a promising source for the production of this enzyme since it is a food-grade organism and its enzyme is rather heat-stable (Greenberg and Mahoney, 1982). These reports suggest that genetic engineering techniques could be used to construct strains capable of overproducing β-galactosidase for commercial purposes. We therefore attempted to clone the genes for β-galactosidase production from the *S. thermophilus* chromosome (Herman and McKay, 1986). From these experiments we were able to isolate a 7.0 kb fragment that contained the genes responsible for producing β-galactosidase. By subcloning and deletion analysis we were further able to localize the gene to about 3 kb. The gene has now been sequenced in our laboratory (unpublished results) and by workers in France (de Vos and Simons, 1988), and we hope to use the information in the construction of vectors applicable in dairy fermentation processes, in the construction of strains that might overproduce the enzyme, and, finally, to begin looking at how lactose utilization is regulated in *S. thermophilus*.

The central role of phospho-β-galactosidase and β-galactosidase in the pathways used for lactose transport and degradation, their properties and distribution in lactic streptococci, as well as data on the cloning, expression, and sequence organization of the genes involved in producing the two enzymes have recently been reviewed by de Vos and Simons (1988). As suggested by these investigators, the lactose genes may well serve as model systems for the study of gene organization, expression, and regulation in lactic streptococci. They also indicated that the introduction of these genes into existing cloning vectors could result in food-grade selection systems for lactic streptococci.

Proteinase Activity

To grow in milk, the mesophilic streptococci are also dependent on their proteinase enzyme system to degrade casein in order to acquire the necessary nitrogenous compounds in the form of amino acids and peptides. This system may also be involved in cheese ripening. Proteinase activity is an unstable property and has now been linked to plasmid DNA in these bacteria. As suggested by Gasson and Davies (1984), the genes for different proteinases could be exchanged between strains and control over the level of their expression should be possible through the application of genetic engineering techniques. This could then lead to strains capable of accelerating the ripening of cheese, either by manipulating the activity of existing enzymes or by the introduction of new enzymes from non-dairy sources. This concept is nearing reality due to the cloning and sequencing of the proteinase genes from *S. cremoris* strains by Venema's group and by de Vos's group in The Netherlands (for review see Kok and Venema, 1988).

Construction of Strains for Accelerated Cheese Ripening

Investigators in the dairy industry have long searched for a method to accelerate the cheese ripening process. Methods that have been attempted all involve increasing the level of enzyme activity or the concentration of enzymes that may be responsible for cheese ripening. Premature cell lysis, such as the use of lysozyme-treated cells or the use of phage for subsequent release of intracellular enzymes has been shown to decrease the intensity of bitterness in cheese and to enhance flavor development (Lowrie, 1977; Law and Wigmore, 1983). Therefore, the isolation of starter culture strains that release their intracellular enzymes at an early stage in cheese production could be a possible method to accelerate cheese ripening. We therefore attempted to isolate mutants of *S. lactis* that would lyse at the cooking temperatures used in Cheddar cheese manufacture. The concept was based on experiments by Shimizu-Kadota in which she was able to isolate thermoinducible mutants from the lysogen *Lactobacillus casei* S-1 (Shimizu-Kadota et al., 1983). By exposing *S. lactis* C2 to a chemical mutagen and then plating to allow colony development at 30°C, followed by replica plating and incubating the plates at 30°C and 40°C, we were able to isolate temperature-sensitive mutants that were able to grow at 30°C but not at 40°C (Feirtag and McKay, 1987a). The growth characteristics of the wild-type *S. lactis* C2 and *S. cremoris* AM2 and two temperature-sensitive mutants, TS11 and TS85, indicated that all four strains grew normally at 30°C. However, *S. lactis* C2 was capable fo growing at 40°C while TS11 and TS85, as well as *S. cremoris* AM2, were unable to grow at this temperature. In addition, in a temperature shift experiment in which the cultures were grown at 30°C for 2.5 hours and then shifted to 40°C, *S. lactis* C2 continued to grow, whereas the growth of *S. cremoris* AM2 was inhibited. In the temperature shift experiments, growth of the temperature-sensitive mutants TS11 and TS85 resembled the growth response exhibited by *S. cremoris* AM2. Since certain *S. cremoris* strains are preferred in cheese manufacture, these results suggest that it may be possible to isolate temperature-sensitive mutants from *S. lactis* which would exhibit the temperature sensitivity normally observed for the desired *S. cremoris* strains. These temperature-sensitive derivatives of *S. lactis* C2 may thus be useful in cheese ripening studies. We were also able to isolate two thermoinducible mutants. These strains grew normally at 30°C but, when shifted to 40°C, upon 1 to 1.5 hours the culture began to lyse as evidenced by a decrease in cell density (Feirtag and McKay, 1987b). In the case of SK11, the decrease in optical density at the cooking temperature used in Cheddar cheese manufacture (38°C-40°C) was found to be due to induction of a prophage. While the nature of the thermolytic response in the *S. lactis* derivatives is unknown, it does not appear to be due to induction of a phage.

The use of thermoinducible lysis mutants in cheese manufacture could have several advantages over the methods currently being attempted to accelerate cheese ripening. First, they could release their intracellular enzymes into the curd matrix at an early stage in the cheesemaking process. Second, they could serve as a "delivery system" for ripening enzymes such as proteinases, peptidases, lipases, esterases, or other enzymes which could improve flavor or quality of the final product. The genes controlling the production of these enzymes could be added to thermoinducible lysis strains through appropriate genetic manipulatory techniques.

Production of Antagonistic Proteins

The ability of dairy streptococci to produce antagonistic substances other than organic acids and hydrogen peroxide is well documented in the literature. These substances include diplococcin, bacteriocins, and nisin. Nisin is by far the best known and most studied inhibitory protein synthesized by dairy streptococci. The possibility of using nisin producing *S. lactis* strains or other inhibitory lactic acid bacteria for food preservation was first suggested by Hirsch et al. (1951). They used nisin producing strains to successfully prevent gas production by *Clostridium* in certain varieties of cheese and were also able to show that nisin producing strains could prevent the growth of *Staphylococcus aureus* in cheese (Hirsch et al., 1951). Nisin is used as a food preservative in some countries and is of considerable economic value. In this regard, nisin produced by starters has been in the human food chain for centuries without apparent negative health effects. This implied safety of nisin suggests an expanding role for nisin producing strains in controlling undesirable organisms that may contaminate milk and milk products. Two excellent reviews of nisin and its potential application are by Lipinska (1977) and Hurst (1981).

Several laboratories have been interested in determining whether the ability of dairy starter cultures to produce antagonistic substances is mediated by plasmid DNA. It is now established that the ability to produce a variety of bacteriocins and diplococcin is linked to these extrachromosomal elements (for review see Klaenhammer, 1988). The possibility that nisin production is plasmid-mediated has been suggested by a number of investigators since the early 1970's. Gasson found that a gene block controlling sucrose metabolism (Suc^+), nisin production (Nip^+), and nisin resistance (Nis^r) was transmissible by a conjugation-like mechanism (Gasson, 1985). In the new host strains a DNA band of about 30 Mdal was observed, but its authenticity as a nisin plasmid could not be established. In agreement with findings by Gasson (1984) and Gonzalez and Kunka (1985), we also observed the conjugal transfer of Suc^+, Nip^+, and Nis^r phenotypes from several donors to various recipients (Steele and McKay, 1986); however, no additional plasmid DNA could be observed in these transconjugants. These results were similar to those reported by earlier investigators (Gasson, 1984; Gonzalez and Kunka, 1985). In addition, we found that the Suc^+, Nip^+, and Nis^r phenotypes could be conjugally transferred to a recipient deficient in its host controlled homologous recombination system, suggesting the genetic determinant could exist as an independent replicon and, if it was being integrated into the chromosome, it did so by a Rec-independent event (Steele and McKay, 1986). The Suc^+, Nip^+, and Nis^r phenotypes also exhibited incompatibility with a number of lactose plasmids from *S. lactis*. This suggests the genetic determinant was not being integrated into the chromosome but did exist as an independent replicon. These results suggest that the *suc, nip* and *nis* genetic determinants in the transconjugants reside either on a currently undetectible plasmid or on DNA which has been integrated into the host chromosome by a Rec-independent event.

When a colony of *S. diacetylactis* WM4 was grown on an agar plate and then overlayed with a sensitive indicator strain, confluent growth of the indicator strain was observed, except around the colony where a clear inhibitory zone was observed (Scherwitz et al., 1983). This indicated that *S. diacetylactis* WM4 was producing some type of antagonistic compound (bacteriocin, Bac$^+$) toward the indicator strain. When WM4 was used as a donor in conjugal matings with a lactose-negative (Lac$^-$) recipient, two types of Lac$^+$ transconjugants were observed. One type was Lac$^+$Bac$^+$, whereas the other type was Lac$^+$Bac$^-$. By examining the plasmid content of these transconjugants, and by curing experiments, it was found that lactose utilization was linked to a 33 Mdal plasmid and the ability to produce Bac was linked to an 88 Mdal plasmid. We then attempted the molecular cloning of the genes responsible for Bac production (Scherwitz-Harmon and McKay, 1987). Digestion of the Bac plasmids with *Bcl*I yielded 17 fragments ranging in size from 1.4 to 16 kb. These fragments were then cloned into the *Bcl*I site on the streptococcal vector pGB301 which codes for erythromycin resistance (Ery^r). After ligation of the *Bcl*I fragments to the vector, the mixture was used to transform *S. lactis* LM0230. After screening hundreds of Ery^r transformants for their ability to produce Bac, a single Ery^{r}Bac$^+$ transformant was isolated. Agarose gel electrophoresis of plasmid DNA isolated from this transformant indicated the presence of a single plasmid, larger than the vector, indicating it contained insert DNA. To determine which of the three *Bcl*I fragments contained the *bac* genes, we randomly examined individual Ery^{r}Bac$^-$ transformants to determine if any contained only the 9.5, 12.1, or 13.4 kb fragments. The presence of any one of these fragments in an Ery^{r}Bac$^-$ transformant would eliminate the possibility of the *bac* genes being on that particular fragment. Plasmid DNA isolated from these transconjugants was digested with *Bcl*I and subjected to agarose gel electrophoresis; three plasmids were found that contained each of the individual fragments (Scherwitz-Harmon and McKay, 1987). These results suggested that a *Bcl*I recognition site existed within the *bac* genes so that cloning of the individual fragments containing a portion of the genes resulted in a Bac$^-$ phenotype. In the recombinant, however, the necessary fragments were present and in the proper orientation to result in a functional gene. A restriction map of the Ery^{r}Bac$^+$ recombinant plasmid was constructed to determine fragment orientation. It was found that the BclI site between the 9.5 and 13.4 kb fragments was probably within the bac genes, since these two fragments were contiguous and mapped together on the 88 Mdal Bac plasmid, whereas the 12.1 kb fragment was not contiguous and mapped elsewhere on the parental 88 Mdal plasmid. Further work indicated that the 9.5 kb fragment coded for resistance to the bacteriocin produced by WM4. The recombinant Ery^{r}Bac$^+$ plasmid also exhibited a greater inhibitory effect than the strain containing only the 88 Mdal parental plasmid. This was assumed to be a gene dosage effect, since the recombinant plasmid appeared to exhibit a higher copy number than did the parental 88 Mdal Bac plasmid. Hopefully it will eventually be possible to construct strains which have increased inhibitory properties active against food spoilage or foodborne pathogens that may occur in our food supply.

Bacteriophage Resistance

One of the major microbiological problems that has faced the dairy fermentation industry over the past 50 years has been the presence of bacteriophage (Daly and Fitzgerald, 1987). From research initiated at Klaenhammer's laboratory, it is now well recognized that the mesophilic lactic streptococci harbor a variety of plasmids that code for mechanisms to defend against this phage attack. These mechanisms, which have recently been reviewed by Klaenhammer (1987), Daly and Fitzgerald (1987), and Sanders (1988), include prevention of phage adsorption, restriction/modification systems, and abortive phage infections. The linkage of plasmid DNA to phage resistance mechanisms, coupled with the observation that many of these plasmids can be conjugally transferred, has led to strain construction strategies for obtaining phage-resistant starters (Klaenhammer, 1987; Sanders, 1988). For example, conjugal strategies have been used successfully to construct fast acid-producing phage resistant transconjugants of S. cremoris and S. lactis for use under commercial conditions (Klaenhammer, 1987; Sanders, 1988). In addition, it has also been suggested that development of phage-resistant strains should include gene and plasmid combinations that provide expression of multiple phage defense mechanisms (Klaenhammer, 1987). This concept has been supported by recent genetic studies showing that combining plasmid-linked mechanisms did indeed confer a higher magnitude of phage resistance (Klaenhammer, 1987; Sanders, 1988). Clearly, a number of genetic approaches now exist for the construction of phage-resistant cultures for commercial purposes. Cloning of the genes responsible for the various phage resistance mechanisms will also facilitate their analysis at the molecular level with respect to expression and regulation. This information, coupled with a better understanding of the phage infection process of lactic streptococci, should aid in strain improvement programs as related to phage resistance.

Exopolymer Production

Exopolymer producing strains of lactic streptococci have been traditionally used in the Scandinavian countries for manufacture of fermented milk products possessing a ropy or mucoid texture. As with many other metabolic properties of lactic streptococci vital for milk fermentation processes, these organisms may also spontaneously lose exopolymer producing ability which results in a loss of the desired viscosity in fermented milks. This observed instability of exopolymer production led some to suggest the involvement of plasmid DNA in its production (McKay, 1983; Macura and Townsley, 1984). Recently it has been shown in a number of laboratories that the ropy phenotype is indeed linked to plasmid DNA (Vedamuthu and Neville, 1986; von Wright and Tynkkynen, 1987; Neve et al., 1988). In addition, Vedamuthu and Neville (1986) demonstrated the conjugal transfer of the mucoid plasmid from a ropy *S. cremoris* strain to several nonmucoid lactic streptococci. The expression of the ropy phenotype was observed in all the transconjugants acquiring the mucoid plasmid. Since the availability of ropy strains is limited, Vedamuthu and Neville (1986) suggested that the conjugal transfer of the mucoid plasmid to other strains could be used to construct additional ropy cultures. These strains could then be used to produce a ropy texture in a variety of fermented milk products. The isolation and characterization of the gene(s) responsible for exopolymer production would also provide further insight with respect to expression and regulation of the ropy phenotype. The latter information could lead to the construction of strains which produce varying degrees of viscosity when propogated in milk.

Vector Construction

Considerable effort has been devoted to the construction of vectors for lactic streptococci. The first vectors used, such as pGB301 (Behnke, 1982) or pSA3 (Dao and Ferretti, 1985), were constructed for other streptococci but were found to function in dairy streptococci. An important observation in the development of plasmid cloning vectors for lactic streptococci was the finding by Vosman and Venema (1983), that a small cryptic plasmid in *S. cremoris* Wg2 could replicate in *Bacillus subtillus*. This led to the construction of a series of lactic streptococcal vectors by Kok et al. (1984) and by de Vos (1987) in The Netherlands, as well as by Gasson and Anderson (1985) in England. An important feature of these vectors was their broad host range since they replicated in lactic acid bacteria, *Bacillus subtillus*, and other Gram-positive bacteria, as well as in *E. coli*. These vectors have now been used to clone genes from dairy streptococci and are also being used to isolate regulatory signals such as promoters, ribosome binding sites, and terminators from this group of bacteria (de Vos, 1987).

Our laboratory has also been interested in constructing vectors for use in lactic acid bacteria. We reported in 1984 that a 40 Mdal plasmid, designated pNP40, from *S. lactis* subsp. *diacetylactis* DRC3 encoded nisin resistance (Nis^r) independent of nisin production (McKay and Baldwin, 1984). This Nis^r phenotype could have several practical implications in the development of cloning vectors applicable to microorganisms used in dairy and food fermentation processes. Since traditional antibiotic selection markers for construction of cloning vectors may be unacceptable due to possible transmission of the drug resistance plasmid in food systems, alternative markers must be used. The Nis^r trait may thus be useful as a selective marker for construction of cloning vehicles applicable to the food fermentation industry. We therefore attempted to clone the genes from pNP40 responsible for Nis^r (Froseth et al., 1988). This 60 kb plasmid from *S. diacetylactis* was restricted with *Eco*RI, and 14 identifiable fragments were generated. This *Eco*RI digest was ligated to the shuttle vector pSA3 which had been linearized with *Eco*RI. The ligation mixture was then used to transform *E. coli*. Recombinant plasmids containing eight of the individual EcoRI fragments of pNP40 were identified, and recombinant plasmids containing the five largest cloned pNP40 EcoRI fragments were isolated from the E. coli transformants and transformed into plasmid-free S. lactis LM0230. Ery^r transformants were selected and all five recombinant plasmids were independently isolated in the LM0230 background. When these Ery^r transformants were examined for Nis^r, it was found that the 7.6 kb EcoRI fragment coded for nis genes. Attempts to further localize the genes revealed that the 7.6 kb EcoRI fragment possessed a single restriction site for XbaI, generating a 2.6 and a 5.0 kb EcoRI-IbaI fragment. Thus, pNPN1 (the recombinant plasmid containing the 7.6 kb insert) contained two XbaI sites, one within the 7.6 kb EcoRI fragment derived from pNP40, and the other within the 10.2 kb EcoRI fragment derived from pSA3. To determine if the Nis^r determinant was located on the 2.6 or the 5.0 kb EcoRI-XbaI fragment, pNPN1 DNA was restricted with XbaI and religated. S. lactis LM0230 protoplasts were transformed with religated DNA and Ery^r transformants were selected and screened for Nis^r. One Nis^r transformant contained a single 11.3 kb plasmid designated pFM040. Restriction analysis indicated it was a self-ligation product of the 11.3 kb XbaI fragment of pNPN1 and contained the 1.6 kb XbaI-EcoRI fragment derived from pNP40. Therefore, the 2.6 kb fragment must encode the Nis^r phenotype of pNP40. A second type of Ery^r transformant plasmid, designated pFM030, was also identified. Transformants containing both pFM040 and pFM030 were cured of the Ery^{r}Nis^r phenotype resulting in the maintenance of only pFM030. Restriction analysis of pFM030 demonstrated that it was a self-ligation product of the 6.5 kb XbaI fragment of pNPN1 composed of the 1.5 kb Xba-EcoRI fragment derived from pSA3 and a 5.0 kb XbaI-EcoRI fragment derived from pNP40. We next attempted to join the 7.6 kb EcoRI fragment encoding Nis^r to DNA fragments containing an S. lactis origin of replication as a first step in construction of a food-grade cloning vector. We attempted to provide and S. lactis origin of replication by using a 5.8 kb cryptic plasmid, designated pJF4628, originating from S. lactis subsp. diacetylactis DRC3 (J.M. Feirtag, M.S. Thesis, Univ. of Minnesota, Minneapolis, 1986). When this plasmid was restricted with EcoRI, two fragments of 3.6 and 2.0 kb were generated. It was assumed that one would contain the origin of replication. pNPN1 and pJF4628 were cleaved with EcoRI and combined to allow fragments to ligate. S. lactis LM0230 protoplasts were then contransformed with this ligation mixture and pSA3 DNA. Ery^r transformants were isolated and evaluated for Nis^r and plasmid content. Two types of Ery^r transformant plasmids were isolated, represented by pFM010 and pFM020. Restriction analysis indicated that pFM010 was composed of the 7.6 kb *Eco*RI fragment and the 2.2 kb *Eco*RI fragment of pJF4628. The result suggested that the 2.2 kb *Eco*RI fragment contained an origin of replication which allowe the 7.6 kb *Eco*RI fragment to exist as an independent replicon. However, further analysis revealed that some Ery^{r}Nis^s transformants, in additon to containing pSA3, also contained a small 3.6 kb plasmid, designated pFM020. Rrestriction analysis of this plasmid indicated it was a self-ligation product of the 3.6 kb *Eco*RI fragment of pJF4628. Since pFM020 existed as an independent replicon, the origin of replication must be located on the 3.6 kb *Eco*RI fragment of pJF4628 since it ws assumed unlikely that pJF4628 would contain more than one origin of replication. It was therefore posible that pFM010 was being replicated by an origin of replication located on the 7.6 kb *Eco*RI fragment of pNP40. To determine if this fragment contained an origin of replication, it was restricted with *Eco*RI and religated. *S. lactis* protoplasts were cotransformed with this ligation mixture and pSA3. Again, Ery^r transformants were evaluated for Nis^r and the Nis^r strain FM011, containing the 7.6 kb plasmid pFM011, was isolated by curing pSA3. Restriction analysis indicated that it was a self-ligation product of the 7.6 kb *Eco*RI fragment. Therefore, pFM011 must encode an origin of replication derived form pNP40, as well as the Nis^r phenotype. A restriction map of pFM011 was constructed to aid in subsequent subcloning, sequencing, and potential use of this plasmid as a cloning vector for lactic streptococci. Unique restriction sites were observed for eight enzymes, which are currently being investigated as potential cloning sites. This plasmid may have potential use as a vector for microorganisms used in dairy and food fermentation processes.

Summary

The genetics of lactic acid bacteria has become a very active area of research, especially for those strains involved in dairy and food fermentation processes. It is now known that in the case of the mesophilic streptococci many of the vital metabolic properties necessary for milk fermentaion processes are encoded by plasmid DNA. This knowledge, coupled with the development of genetic engineering techniques applicable to this group of bacteria, may provide a means of constructing strains useful in dairy and food fermentation processes. There is a continuing need for analysis of the important metabolic traits at the molecular level and for the development of cloning vectors that can be used for microorganisms involved in dairy and food fermentation processes. This knowledge could then lead to the development of strains useful for commercial purposes. As indicated by Gasson and Davies (1984), it is a young field of research and many opportunities exist, not only for application of the genetic studies but also for providing fundamental information about an industrially important group of bacteria.

References

Anderson, D.G. and McKay, L.L. 1984. In vivo cloning of *lac* genes in *Streptococcus lactis* ML3. Appl. Environ. Microbiol. 47: 245.

Behnke, D., Gilmore, M.S. and Ferretti, J. 1982. pGB301 vector plasmid family and its use for molecular cloning in streptococci. In "Microbiology-1982," p. 239. American Society for Microbiology, Washington, D.C.

Blankenship, L.C. and Wells, P.A. 1974. Microbial beta-galactosidase: A survey for neutral pH optimum enzymes. J. Milk Food Technol. 37: 199.

Chopin, M.C., Moillo-Batt, A. and Rouault, A. 1985. Plasmid-mediated UV-protection in *Streptococcus lactis*. FEMS Microbiol. Lett. 26: 243.

Crow, V.L., Davey, G.P., Pearce, L.E. and Thomas, T.D. 1983. Plasmid linkage of the D-tagatose-6-phosphate pathway in *Streptococcus lactis*: Effects of lactose and galactose metabolism. J. Bacteriol. 153: 76.

Dao, M.L. and Ferretti, J.J. 1985. *Streptococcus-Escherichia coli* shuttle vector pSA3 and its use in the cloning of streptococcal genes. Appl. Environ. Microbiol. 49: 115.

Daly, C. and Fitzgerald, G. 1987. Mechanisms of bacteriophage insensitivity in the lactic streptococci. In "Streptococcal Genetics," p. 259. American Society for Microbiology, Washington, D.C.

de Vos, W.M. 1987. Gene cloning and expression in lactic streptococci. FEMS Microbiol. Lett. 48: 281.

de Vos, W.M. and Simons, G. 1988. Molecular cloning of lactose genes in dairy streptococci: The phospho-β-galactosidase and β-galactosidase genes and their expression products. Biochimie 70: 461.

Feirtag, J.M. and McKay, L.L. 1987a. Isolation of *Streptococcus lactis* C2 mutants selected for temperature sensitivity and potential use in cheese manufacture. J. Dairy Sci. 70: 1773.

Feirtag, J.M. and McKay, L.L. 1987b. Thermoinducible lysis of temperature sensitive *Streptococcus cremoris* strains. J. Diary Sci. 70: 1779.

Fitzgerald, G.F. and Gasson, M.J. 1988. *In vivo* gene transfer systems and transposons. Biochimie 70: 489.

Froseth, B.R., Herman, R.E. and McKay, L.L. 1988. Cloning of nisin resistance determinant and replication origin on 7.6 kilobase *Eco*RI fragment of pNP40 from *Streptococcus lactis* subsp. *diacetylactis* DRC3. Appl. Environ. Microbiol. 54: 2136.

Gasson, M.J. 1984. Transfer of sucrose-fermenting ability, nisin resistance and nisin production into *Streptococcus lactis* 712. FEMS Microbiol. Lett. 21: 7.

Gasson, M.J. and Anderson, P.H. 1985. High copy number plasmid vectors for use in lactic streptococci. FEMS Microbiol. Lett. 30: 193.

Gasson, M.J. and Davies, F.L. 1984. The genetics of dairy lactic acid bacteria. In "Advances in the Microbiology and Biochemistry of Cheese and Fermented Milk," p. 99. Elsevier Applied Science Publishers, New York, NY.

Gonzalez, C.F. and Kunka, B.S. 1985. Transfer of sucrose-fermenting ability and nisin production phenotype among lactic streptococci. Appl. Environ. Microbiol. 49: 627.

Greenberg, N.A. and Mahoney, R.R. 1982. Production and characterization of β-galactosidase from *Streptococcus thermophilus*. J. Food Sci. 47: 1824.

Herman, R.E. and McKay, L.L. 1986. Cloning and expression of the β-D-galactosidase gene from *Streptococcus thermophilus* in *Escherichia coli*. Appl. Environ. Microbiol. 52: 45.

Herman, R.E. and McKay, L.L. 1985. Isolation and partial characterization of plasmid DNA from *Streptococcus thermophilus*. Appl. Environ. Microbiol. 50: 1103.

Hirsch, A.E., Grinsted, Chapman, H.R. and Mattick, A.T.R. 1951. A note on the inhibition of an anaerobic sporeformer in Swiss-type cheese by a nisin-producing streptococcus. J. Dairy Res. 18: 205.

Hourigan, J.A. 1984. Nutritional implications of lactose. Austr. J. Dairy Technol. 39: 114.
Hurst, A. 1981. Nisin. Adv. Appl. Microbiol. 27: 85.
Kempler, G.M. and McKay, L.L. 1981. Biochemistry and genetics of citrate utilization in Streptococcus lactis subsp. diacetylactis. J. Dairy Sci. 64: 1527.
Klaenhammer, T.R. 1988. Bacteriocins of lactic acid bacteria. Biochimie 70: 337.
Klaenhammer, T.R. 1987. Plasmid-directed mechanisms for bacteriophage defense in lactic streptococci. FEMS Microbiol. Rev. 46: 313.
Klaenhammer, T.R. and Sanozky, R.B. 1985. Conjugal transfer from Streptococcus lactis ME2 of plasmids encoding phage resistance, nisin resistance, and lactose fermenting ability: Evidence for a high frequency conjugative plasmid responsible for abortive infection of virulent bacteriophage. J. Gen. Microbiol. 131: 1531.
Kok, J., van der Vossen, J.M.B.M. and Venema. 1984. Construction of plasmid cloning vectors for lactic streptococci which also replicate in Bacillus subtilis and Escherichia coli. Appl. Environ. Microbiol. 48: 726.
Kok, J. and Venema. 1988. Genetics of proteinases of lactic acid bacteria. Biochimie 70: 475.
Law, B.A. and Wigmore, A.S. 1983. Accelerated ripening of Cheddar cheese with a commercial proteinase and intracellular enzymes from starter streptococci. J. Dairy Res. 50: 519.
LeBlanc, D.J., Crow, V.L. and Lee, L.N. 1980. Plasmid-mediated carbohydrate catabolic enzymes among strains of Streptococcus lactis. In "Plasmids and Transposons: Environmental Effects and Maintenance Mechanisms," p. 31. Academic Press, NY.
Lipinska, E. 1977. Nisin and its application. In "Antibiotics and Antibiosis in Agriculture," p. 103. Butterworth, London.
Lowrie, R.J. 1977. Influence of lactic streptococci on bitter flavor development in cheese. J. Dairy Sci. 60: 810.
Macura, D. and Townsley, P.M. 1984. Scandinavian ropy milk-identification and characterization of endogenous ropy lactic streptococci and their extracellular excretion. J. Dairy Sci. 67: 735.
McKay, L.L. 1983. Functional properties of plasmids in lactic streptococci. Antonie van Leeuwenhoek 49: 259.

McKay, L.L. 1985. Role of plasmids in starter cultures. In "Bacterial Starter Cultures for Foods, " p. 159. CRC Press, Inc. Boca Raton, FL.

McKay, L.L. and Baldwin, K.A. 1984. A conjugative 40 megadalton plasmid in *Streptococcus lactis* subsp. *diacetylactis* DRC3 is associated with resistance to nisin and bacteriophage. Appl. Environ. Microbiol. 47: 68.

Neve, H., Giels, A. and Teuber, M. 1988. Plasmid-encoded functions of ropy lactic acid streptococci strains from Scandinavian fermented milk. Biochimie 70: 437.

Mercenier, A. and Chassy, B.M. 1988. Strategies for the development of bacterial transformation systems. Biochimie 70: 503.

Rao, M.V. and Datta, S.M. 1978. Lactase activity of microorganisms. Folia Microbiol. 23: 210.

Sanders, M.E. 1988. Phage resistance in lactic acid bacteria. Biochimie 70: 411.

Sanders, M.E. and Nicholson, M.A. 1987. A method for genetic transformation of nonprotoplasted *Streptococcus lactis*. Appl. Environ. Microbiol. 53: 1730.

Scherwitz-Harmon, K. and McKay, L.L. 1987. Restriction enzyme analysis of lactose and bacteriocin plasmids from *Streptococcus lactis* subsp. *diacetylactis* WM4 and cloning of BclI fragments coding for bacteriocin production. Appl. Environ. Microbiol. 53: 1171.

Scherwitz, K.K., Baldwin, K.A. and McKay, L.L. 1983. Plasmid linkage of a bacteriocin-like substance in *Streptococcus lactis* subsp. *diacetylactis* strain WM4: Transferability to *Streptococcus lactis*. Appl. Environ. Microbiol. 45: 1506.

Shimizu-Kadota, M., Sakurai, T. and Tsuchida, N. 1983. Prophage origin of a virulent phage appearing on fermentaions of *Lactobacillus casei* S-1. Appl. Environ. Microbiol. 45: 699.

Smart, J.B., Crow, V.L. and Thomas, T.D. 1985. Lactose hydrolysis in milk and whey using β-galactosidase from *Streptococcus thermophilus*. N.Z.J. Dairy Sci. Technol. 20: 43.

Sorensen, S.G. and Crisan, E.V. 1974. Thermostable lactase from thermophilic fungi. J. Food Sci. 39: 1184.

Steele, J.L. and McKay, L.L. 1986. Partial characterization of the genetic basis for sucrose metabolism and nisin production in *Streptococcus lactis*. Appl. Environ. Microiol. 51: 57.

Vedamuthu, E. and Neville, J.M. 1986. Involvement of a plasmid in production of ropiness (mucoidness) in milk cultures by *Streptococcus cremoris* M5. Appl. Environ. Microbiol. 51: 677.

von Wright, A. and Tynkkynen, S. 1987. Construction of *Streptococcus lactis* subsp. *lactis* strains with a single plasmid associated with mucoid phenotype. Appl. Environ. Microbiol. 53: 1385.

Vosman, B. and Venema, G. 1983. Introduction of a *Streptococcus cremoris* plamid in *Bacillus subtilis*. J. Bacteriol. 156: 920.

Wierzbicki, L.E. and Kosikowski, F.V. 1973. Lactase potential of various microorganisms grown in whey. J. Dairy Sci. 56: 26.

Production of Food Additives and Food Processing Enzymes by Recombinant DNA Technology

Jiunu S. Lai, Jar-How Lee, Shau-Ping Lei, Yun-Long Lin,
Joachim L.Weickmann and Lindley C. Blair
INGENE, Inc.
1545 17th Street
Santa Monica, CA 90404, USA

Introduction

Production of food additives such as antimicrobial compounds, flavor enhancers, low-calorie sweeteners, colors and flavors, nutritional supplements, etc., has been a major challenge for the biotechnology industry since its emergence. The combined use of genetic, protein and fermentation engineering approaches is revolutionizing the traditional food additive market by offering consumers quality products of more versatility, desirability, safety, and economics. It is a powerful tool for 1) large-scale production of rare or high cost materials, 2) improving certain properties of natural products, and 3) eliminating unwanted characteristics of natural products. Here we discuss examples for the application of recombinant DNA technology in the production of an antimicrobial peptide, a low-calorie sweetener, and a group of food processing enzymes.

Cecropin, an Antimicrobial Peptide

From the viewpoint of safety, antimicrobial peptides that could be easily degraded in the digestive tract would be preferred over most chemical preservatives. There are several antimicrobial peptides that have been identified and characterized from natural sources such as seminal plasmin, nisin, magainin, defensin, melittin and cecropin. Melittin is toxic to animals and human beings. Nisin has been approved by several European countries, and was recently approved by the FDA of the U.S.A. as a food additive. However, nisin kills only Gram positive (G(+)) bacteria and is a cyclized peptide with modified amino acids which will be difficult to produce by current recombinant DNA (rDNA) techniques. Defensin has six disulfide linkages, and therefore may be more difficult to produce in a biologically active form by rDNA techniques. Seminal plasmin and magainin have not been sufficiently studied to gather safety or production data for comment. Cecropins appear to be the best candidates for producing by rDNA technology.

Cecropins are a family of small, basic polypeptides found in the hemolymph of *Cecropia* moth pupae during response to bacterial infection (Boman and Steiner, 1981). Only small quantities of these peptides can be isolated from the insect during a particular stage of the life cycle. The ability to produce economically feasible quantities of such material by rDNA technology would allow commercialization as a food and feed additive. We chose to produce cecropin by rDNA techniques for the following reasons:

- It has a wide killing spectrum.
- It is nontoxic to animal cells in culture.
- It is rich in lysine and arginine and can be easily digested in the stomach.
- It is very stable for an extended period of time at pH's as low as 1.5 and as high as 10.0. In addition, the material retains at least ninety percent of its activity after being heated to 100°C for 30 minutes.

We describe here the production of large quantities of cecropins by rDNA technology.

Gene Synthesis and cloning

The cecropin A gene was synthesized in vitro (Lei et al., 1986; Lei and Lai, 1987) and cloned into an *E. coli* expression plasmid pING3. Plasmid pING3 is a pBR322 derivative containing the *araC* gene and part of the *araB* gene from *Salmonella typhimurium* (Lei et al., 1986; Lei and Lai, 1987). The expression of the *araB* gene can be regulated by *araC* protein produced by the *araC* gene on the same plasmid. The araB- cecropin A fusion gene was constructed with a methionine codon at the fusion point and the resulting plasmid was called pCA (Fig. 1). The reasons for constructing a gene encoding a fusion protein instead of simple cecropin peptide are: 1) It is known that small foreign proteins are very unstable in E. coli cells; and 2) A non-fusion cecropin A gene clone may never be usable because once the gene is expressed, the host cells would be killed. The reason for the methionine at the N-terminus of the cecropin A is that active cecropin A can be easily, precisely released upon BrCN cleavage of the fusion protein in vitro. To simplify purification, and to avoid potential problems associated with BrCN, the methionine has been replaced by an aspartate- proline sequence in an alternate construction. The peptide bond between Asp-Pro is cleaved in acidic conditions.

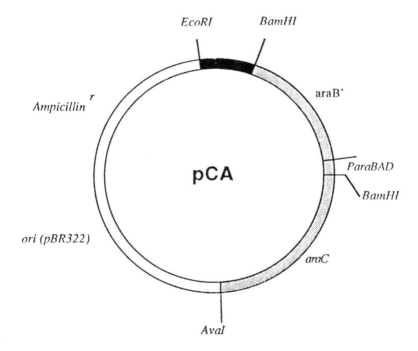

Figure 1. Cecropin expression plasmid PCA

Fermentation
A fed-batch fermentation process is used for production of the fusion protein. Since we find that the induction of fusion protein synthesis

does not affect the growth of E. *coli* cells, we have used the feeding strategy after induction that is shown in Fig. 2. Initially, the recombinant E. *coli* cells are inoculated into the fermentation batch containing minimal medium supplemented with protein hydrolysates. The carbohydrate sources used must be those that cannot cause repression of the *araB* promoter. Cells are grown to an OD_{600} of 10 when L-arabinose is added to turn on the *araB* promoter.

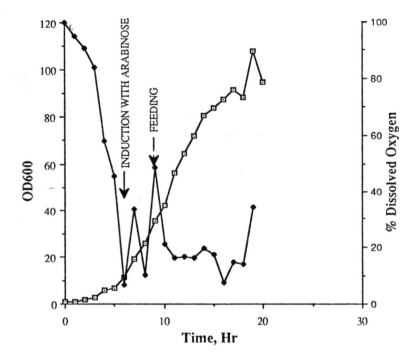

Figure 2. Fermentation profile of E. *coli* producing araB-cecropin fusion protein

Following induction, the fusion protein is synthesized as the culture continues to grow. To prevent the formation of undesirable fermentation products, the dissolved oxygen concentration is kept at approximately 20% of saturation by adjusting agitation and aeration. Feeding the concentrated nutrients to the fermenter commences in response to a sharp increase of percent dissolved oxygen, which indicates the depletion of certain nutrients. At this stage, it is critical that the feeding rate is controlled to maintain a dissolved oxygen level at approximately 20% for maximizing the growth and product yield. This feeding process is programmable and can be automated to warrant a reproducible result. Finally, a high final cell density (>100 OD_{600}) can be achieved with excellent expression of the fusion protein (Fig. 3).

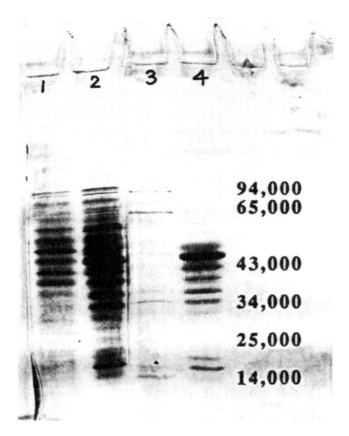

Figure 3. SDS gel electrophoresis analysis of *E. coli* cell extracts showing the quantity of AraB-cecropin fusion protein expression. Lane 1 is the whole cell protein at uninduced condition. Lane 2 is the whole cell protein at induced condition. Lane 3 is molecular weight marker, and Lane 4 is the inclusion body fraction.

Purification

The fusion protein is accumulated inside the E. coli cells in the form of large inclusion bodies, normally two per cell. These refractile inclusions are denser than other cell fractions so they can be pelleted by centrifugation directly after homogenization. Acid is then used to solubilize the inclusions and to release recombinant cecropin from its fusion partner. Further purification by ion exchange chromatography is required to produce food additive quality recombinant cecropin. (Fig. 4).

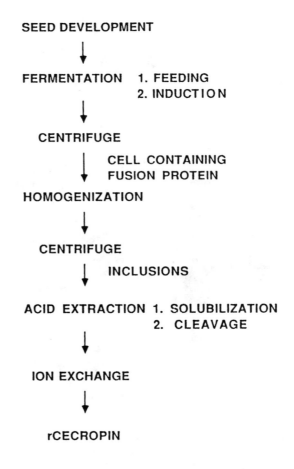

Figure 4. Production scheme of rCecropin

Toxicity Tests

Tests with cecropin were conducted on sheep red blood cells, human B cells, and human fibroblast cell lines. No apparent lysis or damage to the cells was observed at 300 µg/ml concentrations. Acute feeding studies were done with mice. Mice that daily received up to 1.5 grams per kilogram of body weight showed no apparent body-weight change in a three-week period and the enteric bacterial cell count did not differ significantly from the control animals.

Problem and Solutions

Recombinant cecropin A which we have produced is different from cecropin A isolated from the *Cecropia* moth, the natural source of the cecropin A. Natural cecropin A has an amide group at the C-terminus, whereas the recombinant cecropin A has a carboxyl at the C-terminus. When we compared the activity of both materials against different microorganisms, we found that the recombinant cecropin A with a carboxyl group at the C-terminal retains its activity against G(-) bacteria but has reduced activity against G(+) bacteria. To overcome this deficiency, we can modify a variant cecropin A protein *in vitro* and recover its ability to kill G(+) bacteria.

The variant cecropin A is encoded by a gene that has an ATG codon inserted at the 3'-end of the cecropin A gene to facilitate the modification. The C-terminal methionine encoded by this ATG forms a homoserinelactone after BrCN/Formic acid treatment. Then an amine or ethylenediamine adduct can be added through the lactone form (activated carboxyl acid) (Fig. 5). We have named this ethylenediamine modified cecropin, cecromycin. A list of bacteria sensitive to cecromycin is shown in Table 1.

An alternate approach for achieving a wide killing spectrum may be to mix unmodified recombinant cecropin A with nisin. This combination may be ideal as an antibacterial food additive since cecropin A kills G (-) and nisin kills G (+) bacteria (Table 2).

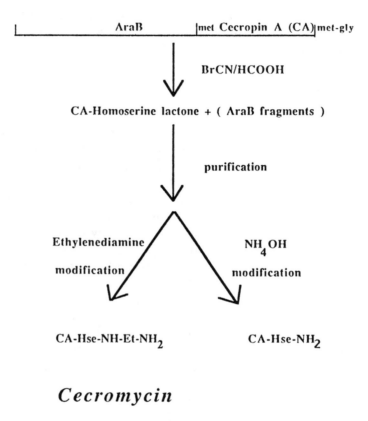

Figure 5. Modification of rCecropin to cecromycin

In conclusion, we have successfully made, through genetic engineering techniques, some antimicrobial peptides with various killing spectra. A primary use of these peptides could be as novel, safer food and animal feed preservatives that do not pose the threats to human health of some chemical preservatives presently in use.

Table 1. Antimicrobial spectrum of Cecromycin

G(-)	G(+)
Bacteroide fragilis	Bacillus megaterium
Enterobacter aerogenes	Bacillus subtilus*
Enterobacter cloacae	Clostridium perfringens
Erwinia chrysanthemi	Micrococcus luteus*
Escherichia coli Proionibacteria acnes	
Klebsiella oxytoca	Staphylococcus aureus*
Klebsiella Pneumoniae	Staphylococcus pyogenes
Neisseria flavescens	Streptococcus agalactiae
Proteus mirabilis	Streptococcus pneumoniae
Proteus rettgeri	
Proteus stuartii	
Proteus vulgaris	
Pseudomonas aeruginosa	
Pseudomonas cepacia	
Rhizobium meliloti	
Salmonella typhimurium	
Salmonella sonnei	
Serratia marcescens	
Shigella entaritidis	
Xanthomonas campestris	
Xenohabdus nemntophillus	

* Minimum inhibitory concentration varies from strain to strain.

Thaumatin, a sweet protein

Thaumatin is an intensely sweet-tasting protein which is about 5,000 times sweeter than sucrose on a weight basis. It is nontoxic, low-caloric and noncaroigenic, and is an excellent candidate for development as a sugar substitute. Thaumatin can also be used as a flavor potentiator at a sub-threshold sweetness level.

The natural source of thaumatin is the arils of the fruit of the African shrub *Thaumatococcus daniellii* Benth. The fruit of *T. daniellii* has traditionally been used as a sweetening substance for sour fruit, corn bread and palm wine in west Africa, and was first described in the Western literature in 1885. It was not until early 1970 that the sweet component in the fruit was identified to be a protein. Currently, the only economical agricultural sources for thaumatin are along the edge of the rain forest belt in certain regions of West Africa. The purified thaumatin is sold under the trade name of Talin by Tate and Lyle, PLC. of England and has been approved as a food additive in many countries including the United Kingdom, Japan, and Australia. However, the supply of thaumatin is difficult to control and the cost of production is still high.

Table 2. Sensitivity of bacteria to Nisin

Organism	Sensitivity to Nisin
G(+)	
Bacillus	+
Clostridium	+
Corynebacterium	+
Pneumococcus type I	+
Streptococci groups A, B, N	+
Staphylococcus aureus	+
G(-)	
Escherichia coli	-
Salmonella typhi	-
Shigella shigae	-
Pseudomonas	-

Since the mid 1970's, thaumatin has been extensively studied. It was found that natural thaumatin consists of at least five variant proteins, two of which appear to be major species. They have similar amino acid composition, sweetness intensity, and molecular weight. But their slightly different charge characteristics allow separation by ion-exchange chromatography on sulfopropyl-sephadex column. The protein sequence of the most abundant thaumatin variant (thaumatin I) has been determined directly (Iyengar et al., 1979). It is a single unmodified peptide chain consisting of 207 amino acids and containing eight disulfide bonds. Disruption of even one disulfide bond in thaumatin destroys its sweetness.

The complementary DNA (cDNA) derived from thaumatin messenger RNA has been cloned and its nucleotide sequence determined (Edens et al., 1982). The protein sequence deduced from the nucleotide sequence differs from the published thaumatin I sequence by five amino acids and was presumed to be that of the other major variant (thaumatin II). The thaumatin II sequence deduced from the cDNA sequence also contains a 22 amino acid long N-terminal pre-sequence (secretion signal sequence) and a 6 amino acid long C-terminal pro-sequence. Both the pre- and pro- sequences are removed from mature thaumatin. Southern blot analysis of *T. daniellii* genomic DNA using cloned thaumatin II cDNA as a probe indicates the existence of a multiple thaumatin gene family.

Having a thaumatin gene enables one to produce thaumatin in microorganisms by fermentation, thus bypassing the problems associated with the current agricultural production method. Earlier attempts to express the cloned thaumatin II cDNA gene in either *Escherichia coli* or *Kluyveromyces* lactis have met with only limited success. Here we describe our approaches to producing biologically active thaumatin from the yeast, *Saccharomyces cerevisiae* (Lee et al., 1988).

Thaumatin protein sequences and their gene syntheses
Because of the discrepancy between the two published sequences, we decided to determine the protein sequences of the two major variants as a preface to designing a synthetic gene (Lee et al., 1988). The two major species of plant thaumatin were identified and separated from commercial thaumatin by SP ion-exchange chromatography. Both proteins were carboxymethylated, purified by FPLC chromatography on a Mono-Q column, and digested with TPCK-trypsin. The peptide fragments were separated by reverse-phase HPLC chromatography on a C18 column, and the sequence of each was determined. In cases where two peptides copurified sequence determination was accomplished by comparing to the published sequences. The sequences of the two thaumatin variants differed by a single residue at position 46 (asparagine vs. lysine). Interestingly, both of them differ from published sequences (Table 3). These two new sequences were designated thaumatins A and B to avoid confusion with the published sequences (Lee et al., 1988).

To express thaumatin as a heterologous product from microorganisms, a synthetic gene coding for the published thaumatin I sequence was constructed (Lee et al., 1988). Genes coding for the thaumatin A and B sequences were derived from the thaumatin I sequence by *in vitro* mutagenesis. An ATG initiation codon was included at the 5' end and two consecutive termination codons were placed at the 3' end of the genes. Yeast preferred codons were used in the thaumatin genes in order to obtain efficient expression of these genes in yeast. DNA sequences encoding the N-terminal secretion signal and the C-terminal pro-sequence were omitted from these constructions.

Table 3. Comparison of amino acid sequence differences between four thaumatin derivatives

Thaumatin	Amio Acid Position				
	46	63	67	76	113
I	Asn	Ser	Lys	Arg	Asn
II	Lys	Arg	Arg	Gln	Asp
A	Asn	Ser	Lys	Arg	Asp
B	Lys	Ser	Lys	Arg	Asp

Expression of thaumatin genes in yeast

The strong yeast promoter for 3-phosphoglycerate kinase (PGK) was used to direct the expression of the thaumatin A, B, and I genes. The yeast PGK terminator was added to the 3' end of the thaumatin genes to provide efficient transcription termination and polyadenylation signals for thaumatin gene transcription. These thaumatin expression cassettes were cloned into a high copy number *E. coli*- yeast shuttle vector, pJDB209. These three expression vectors were transformed into *leu2*⁻ yeast strains and Leu+ transformants were selected (Lee et al., 1988). At least 10% of the protein of the total cellular extract in the Leu$^+$ transformants was found to be thaumatin, as judged by a dense band of the correct molecular weight on a SDS polyacrylamide gel. This band cross-reacted with thaumatin-specific antibodies. The yeast produced thaumatins were extracted from the insoluble fraction of the yeast extract with 6M urea and purified to homogeneity. This thaumatin is not sweet (Lee et al., 1988). Since reduction of the thaumatin disulfides results in protein precipitation and sweetness loss, the reducing nature of the intracellular milieu of yeast may provide one reason why soluble, sweet thaumatin is not found inside these cells.

In order to produce sweet thaumatin, a procedure was developed at INGENE which could refold reduced and denatured plant thaumatin to 25% efficiency. Thaumatins A and B purified from yeast extract can also be refolded to a sweet conformation using our refolding procedure. However, thaumatin I purified from yeast extract did not fold into the sweet conformation after many repeated attempts. Since thaumatin I differs from thaumatin A and B at position 113 of their amino acid sequences (Table 1), we suspect that the amino acid assignment at position 113 of thaumatin I was incorrect (Lee et al., 1988).

Secretion of biologically active (sweet) thaumatin from yeast
Since thaumatin is a secreted protein in the plant, secretion of thaumatin was tested in yeast to see whether thaumatin with the sweet conformation can be produced. A DNA sequence coding for the yeast invertase (*SUC2*) secretion signal was added to the 5' end of the thaumatin I, A and B genes on the expression vectors. The Leu[+] transformants secreted thaumatins A and B into the medium. Secreted thaumatins A and B, purified from culture media, are sweet (Weickmann et al., 1988). In contrast to the A and B genes, the thaumatin I gene does not enable secretion of detectable thaumatin I into the culture media. Furthermore, it is not found inside the cells. We believe that thaumatins A and B properly fold to a sweet form during the secretion process, but that thaumatin I remains unfolded during secretion and may be degraded.

The ability to secrete folded thaumatin from yeast provides a useful alternative to refolding thaumatin extracted from yeast because the downstream processing after fermentation is dramatically reduced. In turn, the ability to economically produce thaumatin by microbial fermentation is preferable to harvesting it from plants where the supply may be susceptible to adverse (e.g. climatic, economic, or political) situations that are beyond the control of the supplier.

Pectic Enzymes for Food Processing

The use of enzymes can make food processing more efficient or make certain products more desirable. However, enzymes used in food processing are usually mixtures of related proteins that co-exist in the starting material. The reason for using only partially purified enzymes is to eliminate the expense of purifying a specific enzyme from a group of proteins which have similar properties. This makes it difficult to control the products of processing through enzyme reactions.

Pectic enzymes are a group of enzymes that can degrade the pectic substances in the plant cell wall (Fig. 6). "Pectinase" has been used as a general term for these enzymes which are widely used in the fruit juice and wine industries to help to improve products and lower production costs in processes such as: 1) accelerating juice and wine clarification, 2) reducing the viscosity of the final juice concentrate, and 3) increasing the efficiency of juice extraction and filtration. Commercial "pectinase" preparations are mixtures of extracellular pectic enzymes, proteases and cellulases. The enzyme mixtures produced from these organisms may show batch to batch variations depending on the strain used or growth media used for their production. Compounding the problem of processing is the fact that different fruits produced in different seasons may have differences in their cell wall compositions and thus specific formulations of pectinase may be required for economic and efficient processing of different batches of fruits.

In order to obtain desired properties of food products and produce them consistently for the purpose of quality control, manufacturers may prefer a customized combination of these enzymes to suit their own purposes. By using genetic engineering methods, one can separate each pectic enzyme by cloning a single gene and expressing it individually. Each enzyme should have different requirements for activity regarding optimum temperature range, optimum pH range, and substrate specificity. Besides custom mixing of different properties enzymes for the different products, one can also manipulate the protein by genetic engineering or protein engineering to get more desirable properties.

Gene cloning and expression
We have cloned the pectic enzyme genes for polygalacturonase and three pectate lyase and expressed them individually in *E. coli* (Lei et al., 1987; Lei et al., 1988; Lee et al., 1988) (Fig.7). When the genes were placed under control of the *araB* promoter, they were efficiently expressed and produced large quantities of pectinases. The natural signal sequence of each protein effects secretion of functional protein out of the *E. coli* cells, allowing a simple purification of the protein from the culture supernatant (Lei et al., 1988) (Fig. 8).

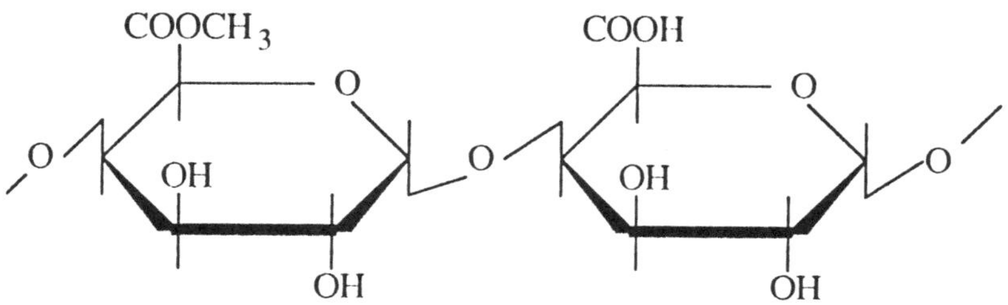

PECTIC ENZYMES

1. Lyases 
   - a. endo-pectate lyase
   - b. endo-pectin lyase
   - c. exo-pectate lyase
   - d. oligo-galacturonate lyase

2. Hydrolases
   - a. endo-polygalacturonase
   - b. exo-polygalacturonase
   - c. oligo-galacturonase
   - d. pectinases

Figure 6. Classification of pectic enzymes according to the action pattern

The pectic enzymes produced by this method should be free of contaminating proteins that may affect the quality of processing the final fruit products, such as juice, wine, etc. We believe that specific mixtures of the pectic enzymes produced by the combined use of genetic, fermentation, and protein technologies will enhance processing efficiency in the food industry.

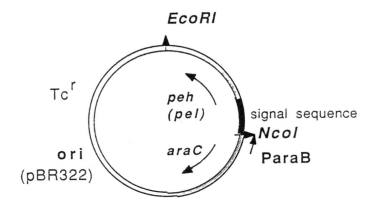

Figure 7. Pectic enzyme expression plasmid

Concluding Remarks

In this article we have described specific examples of products that may be useful to the food industry: cecropin, a natural anti-microbial peptide; pectic enzymes, which are used to break down complex carbohydrates in fruit juices and wines; and thaumatin, an intensely sweet protein, already in use as a low calorie sweetener.

Through the use of recombinant DNA technology we have expressed these proteins in microorganisms as heterologous gene products. Further refinement of our existing technology should lead to economical production of these products for use as food additives or in food processing. Based on this work we believe that novel products, as well as improved versions of existing products, can be developed for the food industry with the aid of this new technology. Following the lead of the pharmaceutical industry, this general approach will soon be widespread in various aspects of the food industry.

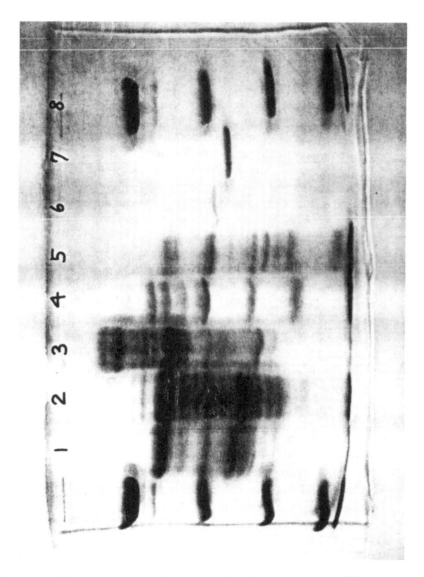

Figure 8. SDS gel electrophoresis analysis of pectic enzymes from different sources. Lane 1: Pectinase from Cooper. Lane 2: Pectinex from Novo. Lane 3: Pectinase from Sigma (*A. niger*). Lane 4: Pectinase from Sigma (*Rhizopus*). Lane 5: Pectolyase from Sigma (*A. niger*). Lane 6: Pectolyase from INGENE. Lane 7: Polygalacturonase from INGENE. Lane 8: Molecular weight standards.

References

Boman, H.G. and Steiner, H. 1981. Humoral Immunity in Cecropia Pupae. Current Topics in Microbiology and Immunology 94/95:75.

Edens, L., Heslinga, L., Klok, R. Ledeboer, A.M., Maat. J., Toonen, M.Y., Visser, C. and Verrips, C.t. 1982. Cloning of the cDNA Encoding the Sweet Tasting Plant Protein Thaumatin and its Expression in *Escherichia coli*, Gene, 18:1.

Iyengar, R.B., Smits, P.S., van der Ouderaa. F., van der Wel, H., van Brouwershaven J., Ravestein., P., Richters, G. and van Wasenaar, P.D. 1979. The Complete Amino Acid Sequence of the Sweet Protein Thaumatin I, Eur. J. Biochem, 96:193.

Lee, J.H., Lai, J.S. and Weickmann, J. 1986. Production of food additives by genetically engineered microorganisms. The World biotechnology Report. Vol 2, P 63 part I. The Proceedings of the Conference held in San Francisco. Online International Inc. London and New York.

Lee, J.H., Weickmann, J.L., Koduri, R.K., Ghosh-Dastidar, P., Saito, K., Blair, L.C., Date, T., Lai, J.S., Hollenberg, S.M., and Kendall, R.L. 1988. Expression of Synthetic Thaumatin Genes in Yeast. Biochem., 27:5101.

Lei, S.P. and Lai, J.S. 1987. Application of Recombinant DNA Technology in Food Industry. The World biotechnology Report. The Proceedings of the Conference held in Santa Clara. Online International Inc. London and New York., pp. 324-329.

Lei, S.P., Lin, H.C., Heffernan, L. and Wilcox, G. 1985. Cloning of the pectate lyase genes from *Erwinia carotovora* and their expression in *Escherichia coli*. Gene 35:63.

Lei, S.P., Lin, H.C., Wang, S.S., Callaway, J. and Wilcox, G. 1987. Characterization of the *Erwinia carotovora* pel B gene and its product Pectate Lyase. J. Bacteriol 169:4379.

Lei, S.P., Lin, H.C., Wang, S.S. and Wilcox, G. 1988. Characterization of the *Erwinia carotovora* pel A gene and its product Pectate Lyase A. Gene 62:159.

Weickmann, J.L., Lee, J.-H., Blair, L.C., Ghosh-Dastidar, P., and Koduri, R.K. 1988. Exploitation of genetic enginering to produce novel protein sweeteners. In "Progress in Sweeteners" (Grenby, T.H., ed.) In press. Elsevier Applied Science Publishers, Essex, England.

Summary

One of the most exciting arenas for application of biotechnology to food quality is the area of bioprocessing. Biological materials are widely used in food processing and can include: microorganisms for dairy, meat, vegetable, and cereal fermentations; food processing enzymes (proteases, carbohydrases, and lipases); and food additiveS such as proteinaceous sweetners or antimicrobials. Historically, mankind has strived to understand and then control microbial and enzymatic bioprocesses that affect the quality and safety of foods. With this information, processing parameters and environments can be established to optimize the beneficial activities of naturally-occurring microorganisms and enzymes. Research efforts over the past decade have now uncovered the molecular, cellular, and genetic technologies that can be employed to manipulate the microorganisms and enzymes responsible for bioprocessing of foods. Examples discussed in the preceding papers illustrate successful applications of these technologies for microbial, enzyme, and protein engineering as well; they also identify future areas where food bioprocesses may be improved and expanded, or new and novel systems created.

Todd Klaenhammer
Department of Food Science
North Carolina State University
Raleigh, NC, USA